T0328946

Logistics outsourcing in the food processing industry

Logistics outsourcing in the food processing industry

A study in the Netherlands and Taiwan

Hsin-I Hsiao

International chains and network series – Volume 8

Wageningen Academic
P u b l i s h e r s

ISBN 978-90-8686-111-8
ISSN 1874-7663

First published, 2009

© Wageningen Academic Publishers
The Netherlands, 2009

Dedicated to my papa, mama and Chih-Sung

Preface and acknowledgements

By doing research on the management of food industry, I learned how to look at this world from a social science point of view. The process was long but joyful.

This book would have never been accomplished without the help of many people.

Firstly, I gratefully acknowledge Wageningen University and the Ministry of Education in Taiwan for providing financial support for my study.

I would like to thank Onno Omta for encouraging me to commence the dissertation research process. I am grateful for his patience and knowledge in guiding me to become an independent researcher. Jack van der Vorst has provided me with invaluable guidance during the research and tremendous mental support. I thank you for your daily support that did not waiver even when you became a professor. I am also extremely grateful to Onno Omta and Jack van der Vorst for making my research financially possible. I would also like to thank Ron Kemp for his meticulous guidance on scientific and statistical issues. The meetings every Wednesday were challenging and inspiring. I benefit a lot from your knowledge, which will be very helpful for my academic career.

I would like to thank the members of my Ph.D. committee for their careful revision of my thesis. Their criticism and constructive suggestions helped me to complete the thesis and it has benefited from their advice and comments.

Part of my research took place in Taiwan. I would like to thank Prof. Tzong-Ru Lee and Prof. Zeng-Ming Zhang for their academic support. I also appreciate the time and effort of Wen-I Huang, Franco Chung and all the companies that I interviewed. I am also grateful to Michelle and Maarten for translating the English summary into Dutch.

I would like to thank all the colleagues at the Business Administration Group, with whom I have discussed the various aspects of conducting the research for this dissertation. In particular, I would like to thank Jacques Trienekens for his supervision and valuable comments in my first year. The discussions with Jos Bijman have also particularly influenced the progression of my research and I would like to thank him for his patiently listening to my torrent of words from time to time. I would also like to thank his family for the friendship they have offered. I would also like to express my warmest thanks to Emiel Wubben with whom I have had many fruitful discussions. Moreover, I would also like to thank Harry Bremmers, and Geoffrey Hagalaar for their guidance. My deepest gratitude also goes to Anna, Dries, Douwe-Frits, Eli, Frances, Guang-Qian, Hans, Ina, Jiqin, Jaime, Leonie, Maarten, Mark, Mersiha, Nel, Paul, Perter, Rannia, Zhen, Willem, Wim; and my formal colleagues, Hualiang, Derk-Jan, Wijnand, Victor, Joanna, and Janneke. Thank you for your help, I enjoyed our academic and cultural discussions particularly during the borrel time.

My special appreciation goes to Pong-Ying Yang, Nico and the Lam family. The researching process was long and sometimes made me feel extremely lonely as a foreign student. I gratefully acknowledge their friendship and hospitality. My special thanks also go to Mr. I-Sung Yang, in Taipei Representative Office in the Belgium for his help during the initial years of my study in the Netherlands.

The support provided by my family has been truly indispensable throughout my study. I would like to thank my parents, sisters and brothers for their endless belief in me throughout my school years and for looking after my little girl, Tina when I was abroad.

Finally, I dedicate this work to my love, Sam Chih-Sung Yang for his endless support and understanding.

Wageningen, December 2008

(Lilly) Hsin-I Hsiao

Table of contents

Chapter 1. Introduction

Logistics management is not a new concept. Throughout the history of mankind wars have been won and lost due to logistics strengths and capabilities or the lack of them. For example, it has been argued that the defeat of the British in the American War of Independence can largely be attributed to a logistics failure. The British Army in America depended almost entirely upon Britain for supplies, but the vital supplies were inadequate to equip and feed the troops overseas until 1781 (Bowler, 1975; Christopher, 2005).

Logistics management can provide a major source of competitive advantage for business organisations (Christopher, 2005). For example, in many industries, logistics costs represent such a significant proportion of total costs that it is possible to make major cost reductions by fundamentally re-designing logistics processes. In addition, customers in all industries are seeking greater responsiveness and reliability from suppliers and they are looking for reduced lead-times, just-in-time delivery and value-added services. It is also possible that superior services can be enhanced through inventory management. However, as the competitive context of business continues to change, traditional methods of managing logistics flows might no longer be valid for ensuring the firm's survival (Bolumole, 2001). Our research interest began by observing the changing competitive environment in the food processing industry, recent development of logistics service providers, and the recent growth of logistics outsourcing. We asked ourselves how food processing companies could keep or increase their competitive advantage by collaborating with logistics service providers (LSPs) to outsource their logistics activities.

Changing competitive environment in the food processing industry

In the last twenty years international trade and foreign production has increased in food industry (Hsiao *et al.*, 2008; Traill and Pitts, 1998). This means that global competition has increasingly intensified, particularly for those firms which previously relied on national regulations to protect them from international competition. In the downstream of food processing, we also observe some dramatic changes. Food retailers have become more concentrated and powerful. Already, today, many retailers are increasingly focusing on their own private labels which will impact the distribution channels and/or product range of food processors. In addition, consumer demand is moving away from commodity products towards more finely differentiated, high quality, value-added products. There is increased concern amongst consumers about the wider non-economic aspects of food consumption; for example, food safety, the environment (e.g. biodegradable packaging) and animal welfare. All these trends are forcing food processors to innovate to find better and more flexible ways to cope with the less stable demand for their products.

New logistics services emerging in the outsourcing market

Logistics outsourcing involves the use of external logistics companies (third-party) to perform activities that have traditionally been performed within an organisation (Bagchi and Virum, 1996; Berglund *et al.*, 1999; Londe and Cooper, 1998; Sink and Langley, 1997). Development of international logistics providers reveals that there has been three main waves of logistics service services emerging in the outsourcing market (Berglund *et al.*, 1999; Carbone and Stone, 2005). In the early 1980s, traditional logistics services firstly emerged; transportation or warehousing services are some of these examples. In the early 1990s, the traditional LSPs began providing value added services through acquisition of specialist capabilities, such as refrigerated transport specialists or assembly specialists. The third wave dates from the late 1990s when a number of players from the areas of information technology, management consultancy and financial services started working together with the LSPs from the first and second wave. The new services called the 'supply chain solution,' also known as fourth-party logistics (4PLs) was introduced in the market because this new LSPs can lead traditional LSPs (3PLs) to supply services to customers (Carbone and Stone, 2005; Hertz and Alfredsson, 2003). Logistics outsourcing is growing in importance worldwide. According to Capgemini (2007), more than 70 percent of the companies in Western Europe and USA have outsourcing experience in transportation or warehousing activities. In addition, we have also seen that the outsourcing trend evolves from basic transportation activities to full logistics network control (Capgemini, 2001, 2002, 2003, 2005, 2007).

Problem statement

In brief, we have seen that the changing competitive environment in the food business has caused many logistical challenges for the food processing industry. To make things more complicate, the food processing industry has some special characteristics that other industries rarely have. For example, seasonality in material production, requirement for conditioned transportation and storage means, or quality decay, which would make logistical planning and transportation more difficult (Grievink *et al.*, 2002). In recent years, researchers have recognised the relevance of supply chain management and innovations for the agri-food sector (Folkers and Koehorst, 1997; Omta, 2004; Taylor, 2006; Van der Vorst and Beulens, 2002; Van der Vorst *et al.*, 2005; Van Duren and Sparling, 1998; Westgren, 1998). However, food industry literature has paid little attention to logistics outsourcing (Bourlakis and Weightman, 2004). As many industries reconfigure their operations around core competencies through outsourcing to react to the changing environments, we might wonder if food companies should re-examine their firm's position within the supply chain to collaborate with LSPs by outsourcing some or all of their logistics activities (Bourlakis and Weightman, 2004).

Given these issues, literature provides almost no guidelines to identify what logistics activities should be sourced out to which type of LSPs for firms are to outsource. Outsourcing of logistics activities usually consists of four steps: (1) identifying the needs, (2) selection of service providers, (3) implementation and (4) service assessment (Sink and Langley, 1997). Outsourcing can be a painful learning experience for companies (Wilding and Juriado, 2004). Sometimes, companies do not renew their contract with LSPs because the goals are not realised or the required service level is not achieved. Hence, making the right outsourcing decision or finding the right LSP is an important issue. Thus, we are particularly interested in the first step of outsourcing i.e. identifying the need and building guidelines for food processors. In order to explore this key issue, the objective of this book is:

> *to analyse how food processors determine their logistics outsourcing need and to analyse how logistics outsourcing influences logistics performance.*

This research will be based on data of Dutch and Taiwanese companies. We have three reasons for this research setup. First, Taiwan is trying to become an international logistics and distribution hub in Asia-Pacific region. The Netherlands is known internationally as the logistics and distribution hub of Europe. Second, both countries are comparable in the sense that they have limited natural resources and land. Third, Dutch agriculture and food processing is famous worldwide; and agriculture is also an important industry for Taiwan. Therefore, it is interesting to compare the two countries where it is expected that Taiwan can learn from the Dutch examples.

In order to realise the objective, four empirical studies are carried out. Let us now introduce the main research questions related to these studies.

Outsourcing decision making framework

Chapter 2 aims at developing a logistics outsourcing decision-making framework, which means identifying the main constructs and the relation between those constructs. The research design in this chapter comprises three stages. First, a literature review is undertaken to study outsourcing theories and to identify outsourceable logistics activities. Successively, exploratory case studies are undertaken to verify the factors and, possibly, identify other relevant factors that are not mentioned in literature. These two stages result in a preliminary decision-making framework. Finally, an exploratory survey is undertaken in the Netherlands to assess the importance of each of the identified factors. Therefore, Chapter 2 aims to answer following question:

> *Research question 1*
>
> *RQ 1 What kind of logistics activities can be outsourced by food processors? (1a) and what decision criteria are considered when outsourcing logistics activities? (1b)*

Outsourcing of level of logistics activities

Chapter 2 presents a decision-making framework, with relevant constructs which influence the outsourcing decision and identifies four levels of logistics activities that can be outsourced. In Chapter 3 we research the relationship between the outsourcing decision at each of these levels with these relevant constructs, more in particular asset specificity, performance measurement uncertainty, core closeness and supply chain complexity. Four propositions are formulated and tested using survey data of logistics mangers in the food processing industry in the Netherlands and Taiwan. Therefore, Chapter 3 seeks to answer the following question:

> *Research question 2*
>
> *RQ2 What decision-making criteria are considered by food processors when outsourcing a certain level of logistics activities?*

Impact on logistics performance

Chapter 4 aims to investigate the impact of different levels of logistics activities on logistics performance. Outsourcing can be a value-enhancing activity. However, the top benefits for companies of outsourcing logistics decisions are often related to costs-savings (Capgemini, 2005, 2007). Among the outsourcing performance-related studies conducted to date, few empirical studies have reported on service benefits; most report on cost performance (Larson and Kulchitsky, 1999; Lau and Zhang, 2006). This chapter seeks to advance our understanding of the relationship between the outsourcing decision, outsourcing level and a firm's logistics service performance. Two propositions are formulated to discuss the direct impact of outsourcing decision on the perceived service performance, and assess the moderating role that supply chain complexity may play in the proposed relationships. The propositions are tested using survey data of logistics mangers in the food processing industry in the Netherlands and Taiwan. Therefore, we seek answers to the following research question:

> *Research question 3*
>
> *RQ3 What is the impact of logistics outsourcing on service performance?*

Taiwan versus the Netherlands

Chapter 2 through 4 investigate outsourcing determinants and the impact of logistics outsourcing on logistics performance. As the last part of this book, we search for implications for the logistics industry (Chapter 5). Taiwan and the Netherlands are located centrally in their geographical regions, the Asia-Pacific rim and Europe. In terms of economic development and logistics environment, the Netherlands is well in advance of Taiwan. The Netherlands is known internationally as the logistics and distribution hub of Europe. Developing Taiwan as an international logistics and distribution hub has become an important issue in the last few years (CEPD, 2002). With its similarity in terms of land size, location and the importance of the agrifood industry, the logistics sector in Taiwan is following the same trajectory as the Netherlands. This chapter compares the logistics outsourcing practices in Taiwan and the Netherlands. To our knowledge no other comparative studies on logistics outsourcing have been conducted between European and Asian countries so far (Arroyo *et al.*, 2006; Sohail *et al.*, 2006). Questionnaires were mailed to logistics mangers in the food processing industry in the Netherlands and Taiwan. Therefore this chapter intends to answer the following question:

Research question 4

RQ 4 What are the current and expected future development in logistics outsourcing in the Netherlands and Taiwan?

Finally, Chapter 6 will summarise and discuss the results. We will also discuss the theoretical and managerial implications of this research.

Chapter 2. Developing a decision-making framework[1]

2.1 Introduction

This chapter aims to answer the first research question.

Research question 1

RQ 1 What kind of logistics activities can be outsourced by food processors? (1a) and what decision criteria are considered when outsourcing logistics activities? (1b)

The purpose of this chapter is to develop a decision-making framework for outsourcing different levels of logistics activities. This is done by first identifying the factors that determine the outsourcing decision of logistics activities. In Section 2.2, a literature review is presented on the outsourcing theories and approaches to determine relevant factors. Section 2.3 presents the results of exploratory case studies that were undertaken to verify the factors and, possibly, identify or specify other relevant factors that were not mentioned in the literature. Together these two stages resulted in a preliminary decision-making framework described in Section 2.4. An exploratory survey was undertaken in the Netherlands to assess the importance of each of the identified factors for each level. These results are presented in Section 2.5. Finally we give the main conclusions and discuss the findings.

2.2 Literature review

In this section, a literature review is presented on the outsourcing theories and approaches to determine the relevant factors.

2.2.1 Definitions

Logistics outsourcing

Many definitions on logistics outsourcing can be found in literature (Bagchi and Virum, 1996; Berglund *et al.*, 1999; Londe and Cooper, 1998). This book uses a combined definition of Lieb *et al.* (1993) and Londe and Cooper (1998): *'logistics outsourcing is a process that involves the use of external logistics companies to perform activities that have traditionally been performed within an organisation, where the shipper and logistics company enter into an agreement for delivering services at specific costs over some identifiable time horizon.'*

---

[1] This chapter is based on the article: Hsiao, H.I., J.G.A.J. van der Vorst, S.W.F. Omta, 2006. Logistics outsourcing in food supply chain networks: theory and practices. In: Bijman *et al.*, International agri-food chains and networks: management and organization, pp. 135-150, Wageningen Academic Publishers, Wageningen.

Logistics process and four levels of activities

Logistics is defined as the process of planning, implementing and controlling the efficient, cost-effective flow and storage of raw materials, in-process inventory, finished goods, and related information from point of origin to point of consumption for the purpose of conforming to customer requirements (Van Goor *et al.*, 2003). A logistics process consists of any activity or group of activities that takes one or more inputs (human assets, equipment, facilities, information, material), transforms and adds value to them, and then provides output (e.g. logistic services) to one or more customers. Table 2.1 summarises the logistics activities most commonly outsourced by large manufacturers. Accordingly, we decompose a logistics process into four levels:

- 1st level: the first level refers to the execution level of basic activities, such as transportation and warehousing. Table 2.1 shows that activities at this level are outsourced to a large degree.
- 2nd level: the second level refers to value-added activities. In food industry, cutting, mixing or packaging are examples.
- 3rd level: this level refers to the planning and control level. Activities that can be outsourced at this level are inventory management and transportation management. Sub-activities of inventory management are sales forecasting, stock control and event control. Sub-activities of transportation management include route planning and scheduling and event control. Table 2.1 shows that activities at this level are less commonly outsourced than the previous levels.
- 4th level: at the top level of logistics activities is the distribution network design. This is the strategic planning and control level in which decisions are made concerning road carrier selection, location and site analysis and logistics network management. When activities at this level are outsourced, the LSP takes care of the logistics network design and orchestrates the logistics flow of the network (Van der Vorst *et al.*, 2007). So far, few studies have included these activities in the investigation.

2.2.2 Outsourcing approaches

In-house or outsourcing (make or buy) decisions have been investigated from different perspectives due to its multidisciplinary nature. Four major approaches are discussed: (1) transaction costs view; (2) resource-bases view; (3) supply chain management theory; (4) other approaches. Here we discuss these theories in general, including historical development, assumptions and predictions on make-or-buy. Table 2.2 summarises the key findings.

Transaction cost theory (TCT)

The first stream of outsourcing approaches is based on Williamson's transaction cost theory (Williamson, 1985). The concept of 'transaction cost' which drives the governance structure was first developed by Coase (1937). Williamson (1985) made great progress by

Table 2.1. Category of logistics activity and the most commonly outsourced activities during 1996-2004.

Category of logistics activity	Lieb and Randall (2002)[a]	Millen et al. (1997)[b]	Wilding and Juriado (2004)[c]
1st level: execution activities			
Fleet management	22%	53%	51%
Shipment consolidation	33%	42%	-
(ocean) Carrier selection	33%	33%	-
Transport	-	-	74%
Rate negotiation	22%	11%	-
Logistics information systems	29%	22%	-
Warehouse management	36%	47%	-
Storage	-	-	60%
Product returns	11%	33%	-
Order fulfilment	9%	33%	-
Order processing	6%	16%	-
2nd level: value-added activities			
Product assembly and installation	11%	13%	-
Re-labelling & re-packaging	-	-	40%
Final product customisation	-	-	37%
3rd level: planning level			
Inventory replenishment & forecasting	6%	-	-
4th level: strategic planning level			
Road carrier selection and site selection	-	-	-

[a] Sample: 500 largest manufacturers in the US identified by Fortune magazine.
[b] Sample: 500 largest firms in Australia in 1994 identified by Business Review Weekly magazine (excluding financial, banking, real estate and insurance organisations).
[c] Sample: 52 consumer goods companies in European countries.

elaborating and operationalising the concept. For the last 20 years, many research studies on boundary choice have been based on the transaction cost concept in particular with regard to production processes or information systems (Aubert *et al.*, 2004; Hair *et al.*, 1998; Robertson and Gatignon, 1998). Transaction cost theory at its core focuses on the costs of completing transactions by one institutional mode rather than another (Williamson, 1975). The transaction, a transfer of a good or service, is the unit of analysis. The primary assumptions are bounded rationality and opportunism which cause transaction difficulties. The theory and empirical studies in Table 2.2 claim that transactions difficulties and associated cost increase

Table 2.2. Summary of key outsourcing approaches, attributes and predictions on outsourcing.

Reference	Sample	Key independent variable(s)
Transaction cost theory		
Anderson (1998)	159 sales managers in electronic manufacturing industry	Transaction specificity Difficulty of evaluating performance Environmental unpredictability
Robertson and Gatignon (1998)	264 R&D directors across a broad spectrum of US industries	Asset specificity Technological uncertainties Behavioural uncertainties
Aubert et al. (2004)	630 IS executives	Asset specificity Uncertainty Business skills, and technical skills
Resource-based view		
Quinn and Hilmer (1994)	Conceptual research	Core competence
Teng et al. (1995)	118 companies	Perceived discrepancy between desired and actual level of performance
Poppo and Zenger (1998)	152 computer executives	Valuable knowledge and capabilities
Insing and Werle (2000)	Conceptual research	Core competence
Arnold (2000)	Conceptual research	Core competence
Leiblein and Miller (2003)	117 semiconductor manufacturers	Sourcing experience
Safizadeh et al. (2008)	108 financial service companies	Degree of customisation service offered
Supply chain management		
Rao and Young (1994)	Cases in wide range of industries	Logistics complexity (product complexity; market complexity; process complexity; network complexity)
Van Damme and Van Amstel (1996)	Conceptual	Demand fluctuation Frequency of delivery
Wilding and Juriado (2004)	50 consumer goods companies in UK, Germany and France	Operational flexibility Cost reduction Expansion to new markets

Key dependent variable(s)	Key findings
Salesman outsourcing	Transaction specificity and difficulty of evaluating performance are related to the use of in-house sales force.
R&D outsourcing	The greater the specificity of existing assets, the more likely that the firm will develop technology internally rather than establish a technology alliance The greater the ability to measure an innovation's performance increase, the more likely alliances are formed.
IT outsourcing	Uncertainty and measurement problems play a role in the IT outsourcing decision.
Outsourcing	A firm should develop a few well-selected core competencies of significance to customers in which the company can be best-in-world, and strategically outsource many other activities where it cannot be best.
Information system outsourcing	When the quality of general Information System support falls short of expectations, the organisation will exhibit a noticeably stronger tendency to outsource.
Information service	Firms internalise and maintain internally those activities in which their superior capabilities enable efficient production.
Outsourcing	Keep core competence in-house and outsource non-core activities because a core activity is an activity with the potential to yield competitive advantage.
Outsourcing	Only the goods and services which are considered to be core competencies should be produced internally.
Production outsourcing	A firm's past experiences affects firm's vertical boundary choices.
Financial service	The greater the degree of customisation offered by a service process, the more likely that, on average, the process maintains its primary back-office activities in-house.
Handling activities, warehousing activities, transportation activities	Logistics complexity is principally due to large volume and variety of logistic transaction. When logistics complexity increases, the likelihood of outsourcing increases.
Logistics outsourcing	The firm would consider outsourcing the logistics to an external logistics provider if the demand of activities is variable.
Transport additional storage during peak periods, fleet management, relabelling, repacking	Consumer goods companies choose to outsource primarily in order to benefit from the competencies of LSPs (skills and flexibility) and to reduce costs. Avoiding investment seems to be particularly important in this capital intensive sector.

when transactions are characterised by three main attributes: *asset specificity*, *uncertainty* and *frequency of the transaction*.

Resource-based view (RBV)

Discussions on a resource-based view of the firm begin with Wernerfelt's (1984) *A Resource-based View of the Firm*, by analysing firms from the resource side rather than from the product side. Following Wernerfelt's article, Barney (1991) proposes a framework, called the resource-based view (RBV) of the firm to study a firm's internal strengths and weaknesses. Assumptions of the resource-based theory are heterogeneous and immobility. Firm resources are controlled by a firm and that enables the firm to conceive and implement strategies designed to improve its efficiency and effectiveness (Barney, 2007: 133). As the resource-based view of the firm has developed, scholars have started a series of discussions on boundary choices, core competencies and competitive advantages. Table 2.2 presents some of these studies; it identifies two main determinant factors for outsourcing: *core competences* and *the value of human assets* for specific business activities.

Supply chain management theory (SCM)

The term 'supply chain management' was first used by Oliver and Webber in 1982 and was developed from logistics point of view. A number of research studies discuss logistics outsourcing from a supply chain management point of view. Rao and Young (1994) and Van Damme and Van Amstel (1996) suggest that firms would consider outsourcing logistics to an external logistics provider when logistics complexity is high. Wilding and Juriado (2004) observe that cost reduction is the main motivation for logistics outsourcing; and Bolumole (2001) also mentions that firms which outsource for operational and cost-based reasons will tend to restrict LSPs' involvement in the basic logistics functions. Therefore, an outsourcing decision might be influenced by a firm's *supply chain characteristics* (e.g. logistics complexity, demand uncertainty) or *logistics strategy*.

In brief, our literature review suggests outsourcing approaches in three categories. Table 2.2 summarises key outsourcing approaches and attributes. As can be seen from the table, a variety of dependent variables are studied, for example, sales forces, production, information service, R&D or logistics activities. In the next stage, three exploratory cases are carried out to verify these factors and possibly identify other factors that impact the decision to outsource logistics activities in the food industry.

2.3 Case studies

In stage 2, three exploratory case studies were undertaken (Voss *et al.*, 2002). For the purposes of confidentiality, the companies are referred to as Company 1, Company 2, and Company 3. These companies were chosen for a number of reasons. Preliminary interviews with managers in these companies revealed that they had outsourcing experiences; in particular, they were outsourcing different activities. Furthermore, they produced different types of food products. The prime sources of data were semi-structured face-to-face in-depth interviews. In each case, descriptions including company background, outsourcing decisions made for logistics activities, motivations, reason for in-house activities and plans for the future are presented.

Company 1

Founded in 1980, Company 1 began as a milk producer for calves. Now it has become one of the largest veal slaughterhouses in the Netherlands. In 2006, Company 1 had more than 160 employees and had reached a turnover of 25 million euros. The corporate policies were to be professional in veal and animal fodders; next to customer-oriented and cost conscious. The logistics strategies aimed for speedy on-time delivery at low cost.

- *Activities and reasons for outsourcing* Transportation and value-added activities were being outsourced when this research was carried out. Company 1 was situated in the centre of the Netherlands. Every week it processed 5,500 calves into a hundred product varieties which were distributed to its 800 customers in the Netherlands and Southern Europe. Time and food quality were critical for the company. To ensure speed and on-time delivery, it had outsourced cutting and packaging activities to local packaging houses. In addition, a number of local road carriers were also hired to execute transport activities. According to the manager the reasons for transportation outsourcing were the lack of their own vehicles and sufficient skills to operate transport activities at the start of business.
- *Reasons for keeping activities in-house* Warehousing, transportation management and inventory management and other logistics activities were still kept in house. Each for its own reason. Warehousing was not outsourced because meat products are very perishable and only stored in a warehouse for a few days before they are delivered. Regarding transportation management, the factory director explained: 'Hiring a logistics company to control all transportations is not an option for us, because tendering transportation services for each market on our own offers better prices. Moreover, time and effort could not be reduced if we outsource this activity.' As for inventory management, the company had its own logistics department doing the job. Overall, these activities are important because food quality and on-time delivery were the competitive priorities. Company 1 could not afford to lose business if logistics companies had problems.
- *Plans for the future* Company 1 was facing two main challenges: the first was increasing competition in Italian and Spanish markets which had forced them to expand into Eastern Europe markets. Another challenge was the time pressures from customers. Nowadays, more and more customers place orders at the last minute, but still request same delivery performance. Nevertheless, even faced with such difficulties, Company 1 had no plan to outsource other logistics activities in the near future.

Company 2

Company 2, founded in 1911, is a dairy company engaged in dairy drinks, buttermilk and yoghurt production. Main activities are production, sales and marketing of fresh dairy products in the Netherlands. In 2006, there were more than 5000 employees and turnover was 600 million euros. The ambition of Company 2 was to be a provider of high-quality fresh dairy products. Low cost and high flexibility were the main logistics objectives.

- *Activities and reasons for outsourcing* Company 2 was very experienced in logistics outsourcing. The transportation activity had been outsourced for 9 years. Company 2 owns three factories and one distribution centre (DC). Every year each factory produced 80 million milks on average with around 300 to 400 varieties. Although demand is quite stable for the whole year, varying distribution channels had made distribution management very complicated. For example, fresh milks were requested to be distributed directly to retailer shops or DCs every day within a 24-hour lead time, while other milk-based drinks were distributed via Company 2's own DC. To deal with such a complex situation, an LSP was contracted for the outbound transportation activities. Cost reduction and logistics had limited added value for the company.
- *Reasons for keeping activities in-house* Other logistics activities were rarely outsourced, such as warehousing or packaging. The traffic manager explained that 'Warehousing is not outsourced because our production heavily depends on warehousing. In addition, warehousing is a very dedicated activity because it deals with very perishable products. Packaging and mixing are rarely outsourced because recipe and know-how are involved in these activities. Therefore, this activity won't be easily outsourced even it is cheaper when done by outside companies.'
- *Plans for the future* The relationship with the LSP was quite stable. In 2004, in an attempt to cut more costs, Company 2 decided to outsource the route planning (transportation management). When transportation management was outsourced, a logistics company would take over some of the jobs from the traffic manager, such as planning for distribution time for its own logistical efficiency maximisation. However, this new relationship was not successful because the proposed delivery time window was rejected by retailers. Therefore, this activity was taken back in-house again. Lack of flexibility of the LSP was one of the possible explanations for this failure. The traffic manager stated that for the coming three years no other activities would be considered for outsourcing.

Company 3

Company 3 produces non-perishable food products which can be stored at ambient temperatures. The main activities of Company 3 are production, sales and marketing of long-life dairy products, as well as branded fruit juices and fruit-based drinks. In 2006, there were about 5,000 employees with a 600 million euro turnover. The main corporate objective was to manufacture and market dairy and fruit-based products in such a way as to create value that could be sustained in the long run for customers, shareholders, employees and business partners. Logistics, from the company's point of view, did not add much value to this process.

- *Activities and reasons for outsourcing* Outbound transport and transportation management were being outsourced. Company 3 owned three factories and one DC. Each year the three factories produced around 500,000 pallets for its 250 customers in Dutch, Belgian and German markets. Demand for their products fluctuated significantly; for example, juice products sometimes had a high peak demand in summer. To deal with such a situation, a long-life food specialist LSP was contracted to operate both the tactical and operational planning for Company 3. The main reasons for hiring the LSP were its specific competences and assets in terms of ambient food logistics, and cost reduction. 'We had incurred losses over the past few years with increased price competition between retailers,' the logistics manager explained. Therefore cost reductions were necessary. In 1994, a new contract was signed, signalling that the LSP would start taking over transportation management activities for Company 3. Since then, every 3 to 6 months the LSP has presented a distribution schedule to Company 3 on volume, time and distributions by planning its own resources and equipment for efficiency maximisation.
- *Plans for the future* So far, the outsourcing programme has been quite successful and satisfactory; for instance, on-time delivery has increased to 93% and logistics costs have been reduced. Although the price war between supermarkets is still a pressure, Company 3 has no plans at present to outsource other logistics activities.

In brief, the responses from the interview data gathered from the participants exemplify the relationships between activities and outsourcing motivations. Table 2.3 summarises the reasons for both outsourcing and not outsourcing. Most of these findings are compatible with the findings in Stage 1. Based on these findings, we created an outsourcing decision-making framework.

2.4 Construction of the decision-making framework

Literature and cases demonstrate examples of outsourcing considerations. This helps us to construct a preliminary decision-making framework which presents key determinants for logistics outsourcing in food industry, shown in Figure 2.1 Asset specificity and measuring uncertainty are the attributes of TCT; core closeness is the attribute of RBV; supply chain complexity and logistics strategy are based on SCM. Five propositions are then formulated concerning the determinants and its predictions on outsourcing decisions. Performances are assumed to improve if outsourcing decisions are made.

Asset specificity

Our case results suggest including asset specificity into the framework because the cases show that a logistics outsourcing decision is influenced by existing assets such as dedicated facilities or current investments in employees' logistical planning skills, etc. This finding is

Table 2.3. Summary of case results.

Reasons for outsourcing

Example: transportation, cutting, packaging, transportation management
- do not own any transport vehicle since the company's start (*asset specificity-TCT*)
- the activity is less important to company (not the core business-*RBV*)
- logistics not adds a lot of value to company (*low valuable-RBV*)
- cost reduction (*logistics strategy-SCM*)
- complicated logistics requirements, such as demand fluctuation, serving many international customers, etc. (*supply chain complexity-SCM*)
- cost pressure – price war between retailers (*other perspectives*)
- lack of professional knowledge (*RBV*)

Reasons for in-house

Example: warehousing, inventory management, transportation management, distribution network design, etc.
- many years of experiences in finding cheap carriers (*asset specificity-TCT*)
- the activity can be operated by our own logistics department (*asset specificity-TCT*)
- the activity is very dedicated for own products (*asset specificity-TCT*)
- time and effort cannot be reduced (*transaction uncertainty-TCT*)
- outsourcing these activities will damage our core business (*important to core business-RBV*)
- food quality, speed, flexibility were the competitive priorities (*logistics strategy-SCM*)

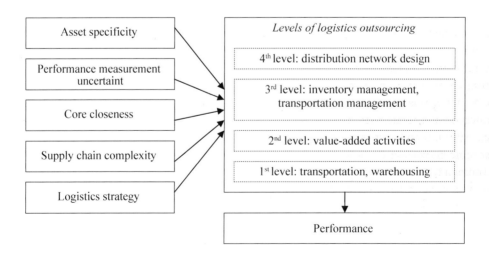

Figure 2.1. A preliminary decision-making framework for levels of logistics outsourcing in the food processing industry.

in accordance with other literature (Aubert *et al.*, 2004; Hair *et al.*, 1998; Robertson and Gatignon, 1998). The argument is as follows: logistics-specific assets involve investments in human and physical capital which will lose value if they are redeployed for other uses. Thus, the following proposition is formulated.

> *Proposition 1: The higher the asset specificity of a specific logistics activity, the less likely that a food processor will outsource this activity than keep it in-house.*

Performance measuring uncertainty

Literature and case results also suggest including performance measurement uncertainty into the framework. One of the in-house reasons is that time and effort cannot be reduced. This time and effort problem might relate to transaction uncertainty, in particular, the behaviour uncertainties. Behavioural uncertainty is the difficulty associated with monitoring the contractual performance of exchange partners. This finding is compatible with previous research that identifies the importance of transaction uncertainty on outsourcing decisions (Poppo and Zenger, 1998; Robertson and Gatignon, 1998). Therefore, we argue that when performances cannot be easily assessed, outsourcing can be 'inefficient' (i.e. less profitable than in-house). The contracting costs are higher when writing an incentive compatible contract under a complex performance assessment. Thus we formulate the following proposition:

> *Proposition 2: The higher the performance measuring uncertainty when outsourcing a logistics activity, the less likely that a food processor will outsource this activity than keep it in-house.*

Core activity or not?

In this chapter, we focus on a logistics activity's 'core closeness' instead of core competence (Franceschini *et al.*, 2003; Rao and Young, 1994), because the core business of food processors is most often research and development, not logistics. Our initial finding from cases indicates that an outsourcing decision might depend on an activity's value to core business (core closeness). This finding is consistent with previous research that discusses the value of human assets on outsourcing decisions (Conner and Prahalad, 1996; Poppo and Zenger, 1998; Teng *et al.*, 1995). Each logistics activity has human assets related to it, indicating a firm's general and specific knowledge on how to do things, for instance, transportation requires driving skills. Therefore, we argue that firms internalise and maintain internally those activities in which their superior capabilities or knowledge enable efficient logistical performance. Accordingly, we formulate another proposition.

> *Proposition 3: The closer a logistics activity is to the core business, the less likely that a food processor will outsource this activity than keep it in-house.*

Supply chain complexity

Our case results indicate that supply chain complexities (for instance, number of products, demand prediction, number of international customers, and distribution channel variety) complicate logistical planning. Complexity causes planning and control problems to firms and influences firms' performances (Milgate, 2001). Literature suggests that when firms want to increase performances, firms can redesign chain structures (Van der Vorst and Beulens, 2002), and shift part of the complexity out-of-house (Wang and Von Tunzelmann, 2000). Therefore, we argue that supply chain complexity is positively related to an outsourcing decision because a supply chain with a high degree of complexity might require an LSP to reduce the managerial complexity. Therefore, we formulate the following proposition:

> *Proposition 4: The higher the supply chain complexity, the more likely a food processor will outsource a logistics activity than keep it in-house.*

Logistics strategy

The cases show that cost reduction is one of the outsourcing considerations, and companies with food quality, speed or flexibility priorities prefer to keep an activity in-house. This result partly fits with previous studies. Bolumole (2001) mentions that outsourcing of basic logistics functions is based on operational and cost-based reasons; Al-Kaabi *et al.* (2007) studied the outsourcing of maintenance, repair and overhaul in the airline industry and concluded that the low cost airlines and new airline entrants preferred to outsource all maintenance, repair and overhaul activities. However, companies with food quality or flexibility priorities prefer to keep an activity in-house because they worry logistics companies have limited knowledge of food quality management or lack of flexibility. Therefore we suggest including logistics strategy into the framework, and we propose:

> *Proposition 5: A food processor with a low cost strategy is more likely to outsource a logistics activity than a food processor with a flexibility or food quality strategy.*

2.5 Survey

The literature review and case results create a conceptual framework, but it doesn't provide quantitative insight into the relevance of each variable for the outsourcing decision of each level of activities. The objective of this stage is to test whether activities at different levels are outsourced for different reasons. This part discusses findings in relation to our third research question.

Research method

An exploratory survey (Forza, 2002) was undertaken at this stage. To analyse the outsourcing decisions, three different activities at different levels were chosen: transportation (1st), packaging (2nd) and transportation management (3rd). These three activities were selected because they were also outsourced in the cases. Company names and addresses were obtained from the Dutch Chamber of Commerce (www.kvk.nl). Companies with more than 40 employees were chosen. A total of 385 questionnaires were sent. After two months, 57 were considered useable and were returned by postal service. A total of 76 responses were received of which 14 had missing data and were judged unusable. The final sample size was 62, resulting in a respond rate of 19%.

Measures and analysis

The variables in the survey include: make-or-buy choices, asset specificity, performance measurement uncertainty, core closeness, logistics strategy and supply chain complexity. Measures of these variables are described in Appendix 1. A t-test was used to compare two population means (the outsourced group and in-house group) on the variables: asset specificity, measuring uncertainty, core closeness and supply chain complexity. The logistic strategy variable was measured by the proportions test.

Results

Table 2.4 presents comparisons between the outsourced and in-house groups on the different activities. The test statistics shows that transportation outsourcing is influenced only by asset specificity ($P<0.001$). The packaging outsourcing is influence not only by asset specificity ($P<0.001$), but also by the complexity caused by the number of products produced ($P<0.05$), demand uncertainty ($P<0.05$) and the average of supply chain complexity ($P<0.05$). Further, transportation management is also outsourced under multiple conditions. Besides asset specificity ($P<0.01$), distribution channel variety ($P<0.1$) and the average of supply chain complexity ($P<0.1$), the transportation management is also influenced by core closeness ($P<0.01$). Logistics strategy does not differ significantly between groups in the three activities, but we observed that cost priority seems higher in the outsource group particularly in the transportation and packaging activities and higher in the in-house group for transportation management. To summarise, the results show that different activities are outsourced for different reasons.

2.6 Discussion and conclusions

Much management literature exists on the outsourcing of production activities or information systems, but there is a gap in the literature regarding the outsourcing of the planning level of logistics activities especially in the food industry. This research takes a first step in bringing

Table 2.4. Comparisons of the outsourced group (G_o) and in-house group (G_{ih}) on the three levels of logistics activities[a,b].

Variables	1st level Transportation			2nd level Packaging			3rd level Transportation management		
	G_o (N=45)	G_{ih} (N=17)		G_o (N=5)	G_{ih} (N=57)		G_o (N=22)	G_{ih} (N=40)	
Asset specificity	3.60	6.94	***	2.60	6.84	***	4.50	5.82	**
Performance measurement uncertainty	6.20	5.50		4.80	5.98		5.82	5.66	
Core closeness	6.16	7.00		7.20	6.81		6.60	7.33	*
Supply chain complexity									
- number of products (SKUs)	4.80	4.90		6.00	4.74	*	4.95	4.78	
- demand uncertainty	4.84	4.44		6.20	4.61	*	5.14	4.51	
- number of international customers	3.50	3.00		3.80	3.32		3.67	3.20	
- distribution channel variety	3.58	3.00		4.20	3.35		4.00	3.10	+
- avg of supply chain complexity	4.33	3.99		5.25	4.15	*	4.59	4.04	+
Logistics strategy									
Cost	51%	47%		60%	49%		32%	53%	
Flexibility	18%	29%		20%	21%		27%	25%	
Food quality	16%	6%		20%	12%		9%	15%	
Others	16%	18%		0%	18%		32%	8%	

[a] $+P<0.1$; $*P<0.05$; $**P<0.01$; $***P<0.001$ (2-tailed).
[b] Outcomes of packaging were also confirmed by Kruskal-Wallis test due to the small number in G_o, results were the same.

outsourcing research in this context to create a conceptual framework for outsourcing different levels of logistics activities. This chapter presents two principle findings:

1. *A logistics outsourcing decision is related to asset specificity, core closeness, and supply chain complexity.* The framework identifies these determinant factors for the food industry. We propose that the lower the current investment by the firm in logistics assets, the higher the likelihood that that activity is outsourced than in-house (proposition 1). The less close the activity to the core business, the higher the likelihood that an activity is outsourced than in-house (proposition 3). Moreover, the greater the supply chain complexity, the higher the likelihood that an activity is outsourced than in-house (proposition 4).

2. *Logistics activities at different levels are outsourced for different reasons. Evaluation of different activities requires insights into three theories- transaction cost, resource-based and supply chain management theory.* There is a growing body of literature discussing the fact that TCE and RBV are complementary - that each theoretical perspective alone cannot fully explain a make-or-buy decision (Arnold, 2000; Holcomb and Hitt, 2007; Madhok, 2002; Poppo and Zenger, 1998). Interestingly, our preliminary findings echo these expectations but add that SCM theory – especially supply chain complexity – should be taken along as well.

To conclude, a conceptual decision-making framework is created for interpreting and analysing outsourcing considerations of different levels of logistics activities. The framework identifies three determinant factors for the food industry: asset specificity, core closeness, and supply chain complexity. In addition, our outsourcing framework indicates that logistics activities at different levels are outsourced for different reasons. Performance measurement uncertainty and the logistics strategy might also have an influence, however this could not be proven via the survey.

There are some limitations in this research. First, the sample cannot be used to generalise about the overall population of the whole food industry. Therefore, more sample data is needed to test the proposed model and provide sufficient evidence. Second, we only tested three logistics activities in the survey. Further research is needed which includes more logistics activities at the successive levels.

Chapter 3. Outsourcing decisions and level of logistics outsourcing[2]

3.1 Introduction

This chapter aims to answer the second research question.

Research question 2

RQ2 What decision-making criteria are considered by food processors when outsourcing a certain level of logistics activities?

The previous chapter has divided outsourceable logistics activities into four levels. These four levels in sequence are: the *1st level* of logistics, referring to the execution level of activities, such as transportation and warehousing activity. The *2nd level* refers to the value-added activities, such as packaging, labelling or mixing, etc. The *3rd level* refers to the planning and control level of logistics activities. Examples are inventory management (referring to sales forecasting and stock control) or transportation management (referring to route planning and scheduling). The *4th level* is often called fourth-party logistics or distribution network design. This is a strategic planning and control level. Examples are location and site selection or road carrier selection. When activities at this level are outsourced, the LSPs take care of the logistics network design and orchestrate the logistics flow of the network (Van der Vorst *et al.*, 2007). Thus, the five activities studied by our research are transportation (1st), packaging (2nd), transportation management (3rd), inventory management (3rd) and distribution network design (4th).

This chapter proceeds as follows: the next section presents a literature review and develops research hypotheses. In Section 3.3, the research design is presented with details on data collection and constructs we apply in this research. In Section 3.4, results are presented. Finally, Section 3.5 discusses the implications of our findings, points out limitations and suggests further research.

3.2 Research framework

A firm can be seen as both a collection of transactions and a bundle of resources. A growing number of studies propose that the decision with respect to the appropriate governance structure rests not just on cost, but also on productivity benefits tied to skills and knowledge, and on the configurations within the firm (Holcomb and Hitt, 2007; Jacobides and Hitt, 2005; Madhok, 2002). In this section, we discuss a firm's make-or-buy decision of a certain

---

[2] This chapter is based on an article accepted for publication in the Journal on Chain and Network Science: Hsiao, H.I., R.G.M. Kemp, J.G.A.J. van der Vorst and S.W.F. Omta, Make-or-buy decisions and level of logistics outsourcing An empirical analysis in the food manufacturing industry.

logistics activity from three perspectives: transportation-based, resource-based and supply chain logistics-based. Based on these perspectives, we formulate hypotheses.

Transaction cost theory: asset specificity and performance measurement uncertainty

Transaction cost economics (TCE) describes firms as a governance structure in which firms and markets are alternative modes of governance. Rather than view the efficient boundaries of the firm in terms of technology (economies of scale and scope), the efficient boundaries can be derived by aligning different transactions with governance structures (firm or market) in a discriminating way (Williamson, 1998). Thus, we use transaction cost economics to evaluate a make-or-buy decision (i.e. firm-or-market).

Transactions, which differ in their attributes, are aligned with governance structures, which differ in their cost and competence. In selecting the 'right' governance mode for a transaction, the transaction costs should be minimised (Williamson, 1998). Two important attributes that influences the transaction costs are asset specificity and uncertainty (David and Han, 2004; Williamson, 1975).

Asset specificity takes a variety of forms: physical assets, human assets, site specificity, dedicated assets, brand name capital, and temporal specificity (Williamson, 1998). Transaction-specific assets involve investments in human and physical capital that cannot be redeployed without losing productive value. These assets may be the specific knowledge or expertise to carry out a certain activity or serve a particular customer through collective learning or accumulated experience in a certain time (human asset specificity) or these assets may be in plant and equipment that are dedicated to producing a specific product or service (physical asset specificity) (Robertson and Gatignon, 1998). Empirical research has provided strong and consistent support for the theorised relationships between transaction-specific investment and governance form (Aubert *et al.*, 1996; Aubert *et al.*, 2004; Dyer, 1997; Grover and Malhotra, 2003; Poppo and Zenger, 1998; Yasuda, 2005). Robertson and Gatingnon (1998) showed a negative relationship between asset specificity and the decision to use R&D alliances in the development of technology. Poppo (1998) finds that the presence of firm-specific assets encourages internalisation. Based on these statements, we formulate the first proposition. When the specificity of existing assets is high, the governance costs of alliances (outsourcing) render them inferior to internal modes (Williamson, 1991). Therefore we propose:

> *Proposition 1: The higher the asset specificity of a specific logistics activity, the less likely that a food processor will outsource this activity than keep it in-house.*

Transaction uncertainty can be divided into three categories (David and Han, 2004): market condition uncertainty; technology uncertainty; and behavioural uncertainty. Behavioural uncertainty is especially important for make-or-buy decisions. Performance measurement uncertainty is one dimension of behavioral uncertainties (David and Han, 2004) which focuses on the difficulty in evaluating performance (Poppo and Zenger, 1998). It refers to the degree of difficulty associated with assessing the performance of transaction partners (Rindfleisch and Heide, 1997). Empirical studies have shown a negative relationship between measurement uncertainty and outsourcing (buy) decisions (Hair *et al.*, 1998; Poppo and Zenger, 1998; Robertson and Gatignon, 1998). In brief, when performances cannot be easily assessed, using markets can be 'inefficient' (i.e. less profitable than in-house), the contracting costs are high when writing an incentive compatible contract under a complex performance assessment. Thus we formulate another proposition:

> *Proposition 2: The higher the performance measuring uncertainty when outsourcing a logistics activity, the less likely that a food processor will outsource this activity than keep it in-house.*

Resource-based view: core closeness

The resource-based view (RBV) asserts that firms gain and sustain competitive advantages by developing valuable resources and capabilities (Barney, 1991). Firms internalise and maintain internally those activities in which their superior capabilities enable efficient production (Poppo and Zenger (1998). Insinga and Werle (2000) divide business activities into four different types according to their potential to yield competitive advantage: commodity activity (readily available), basic activity (needed to be in the business), emerging activity (has the potential to become a competitive differentiator) and key activity (a competitive differentiator, or core activity). The core activities are central to the company successfully serving the needs of potential customers in each market (McIvor *et al.*, 1997). Literatures have suggested that core activities will not be outsourced, because this allows firms to leverage their unique competencies (Insinga and Werle, 2000; Leiblein and Miller, 2003) and offer a long-term competitive advantage (Quinn and Hilmer, 1994). In this chapter, we use core closeness as a decisive criterion to measure a logistics activity because the core business to a food firm usually refers to processing activities or product development and design, but not logistics activities.

Logistical resources include tangible assets (such as trucks or warehouses) and intangible assets (such as knowledge or skills, i.e. 'capability'). Capability is a complex bundle of individual skills, accumulated knowledge exercised through an organisational process that enable firms to co-ordinate activities and make use of their resources (Olavarrieta and Ellinger, 1997). It could also be developed experientially at the firm or plant level (Leiblein and Miller, 2003). A logistics activity is executed or translated by an employee's capabilities. For example, a transportation activity is executed by a truck driver's driving skills; an inventory management activity is executed by employees' ability to predict stock levels through experiences and use of software. These available capabilities also influence the make-or-buy decision. For instance, Argyres (1996) proposed that firms vertically integrate into those activities in which they have greater production experience and/or organisational skills (i.e. 'capabilities') than potential suppliers, and outsource activities in which they have inferior capabilities. In brief, we assert that firms internalise a certain logistic activity in which they have superior capabilities to gain value for firms, i.e. the activity is close to the core business. We propose:

Proposition 3: The closer a logistics activity is to the core business, the less likely that a food processor will outsource this activity than keep it in-house.

Supply chain management: supply chain complexity

In this research, supply chain management theory is viewed from logistics perspectives (Londe and Cooper, 1998; Stevens, 1989; Van der Vorst, 2000). Literature rarely uses the supply chain management approach to evaluate a make-or-buy decision. In 1994, Rao and Young conceptually mentioned that logistics outsourcing decisions might be related to supply chain characteristics, such as product complexity (perishability, size, density), process complexity (time sensitivity, manufacturing cycle), and network complexity (number of trading companies, countries and continents) (Rao and Young, 1994). Other authors also mention that an increase in supply chain complexity could deteriorate delivery performance, so we add supply chain complexity as one of the considerations in the make-or-buy decision (Hsiao *et al.*, 2006; Milgate, 2001).

Complexity refers to the level and type of interactions present in the system (Milgate, 2001). Complexity is viewed as a deterministic component more related to the numerousness and variety in the system. Building on the conceptual definitions of complexity used by Milgate (2001) and Choi (2006), we regard supply chain complexity as associated with the 'number of elements' within the system and the degree to which these elements are 'differentiated' in the logistics concept. In this regard, *supply chain complexity* means the number of elements within the focal company's logistical flow (on bases of supply, production, distribution and demand), and the degree to which these bases are differentiated.

The level of supply chain complexity affects the level of effort, or operational load, required to manage a system (Choi and Krause, 2006). For instance, a large number of suppliers increases the level of coordination needed to improve efficiency of operations. With fewer suppliers, the focal company can implement a more efficient buyer-supplier interface through more cost-effective inventory control. To summarise, supply chain complexity is assumed to be higher for a focal company if its supply, production, distribution and demand bases are in great number or varying to a large degree. Then focal company requires a high level of effort or operational load to manage this system. To ease or transfer such complexity, the focal company might seek to form a logistic alliance with logistics companies.

Proposition 4: The higher the supply chain complexity, the more likely a food processor will outsource a logistics activity than keep it in-house.

Control variables

Three variables are considered as control variables: *geographic location*, *size of firm* and *change of sales growth rate*. These variables can possibly influence an outsourcing decision. However, we don't set an expectation on the relationships between these and the outsourcing decision. Take firm size as an illustration. Even with available funding for internal logistics activities, larger firms could also favour outsourcing because they may have greater bargaining power to lower price (Robertson and Gatignon, 1998).

3.3 Data

The research framework was tested using Dutch and Taiwanese data. The sample frame consists of a mailing list of food manufacturing firms from membership lists of the Dutch Chamber of Commerce (www.ksv.nl) and Taiwan's Industry & Technology Intelligence Service (www. itis.org.tw). Surveys were mailed to logistics managers in firms with at least forty employees. Following Groves et al's (2004) survey methodology, initial mailings were followed by phone calls after two weeks. If necessary, second mailings were carried out. Data was gathered from September 2006 to February 2007. Of the 890 surveys mailed (NL: 385; TW: 505), 66 had incorrect contact information (NL: 57; TW: 9) and were returned by the postal service. A total of 138 responses were received (NL: 76; TW: 62), of which 24 had missing data (NL: 7; TW: 17) and were judged unusable, thus yielding a sample size of 114 (NL: 69; TW: 45) with a response rate of 15% (114/800) (NL: 21%; TW: 9%). The response numbers to the studied variables is close to the recommended rule of thumb for binary logistic regression (Hair *et al.*, 1998).

3.3.1 Measures

The constructs and underlying questionnaire items used in this study are shown in appendix 2. Reliability coefficients are presented as well. Cronbach alphas in most of constructs are well above 0.7, except performance measurement uncertainty. Below, we describe the constructs in more detail.

Make or buy decision

Five logistics activities are identified: transportation, packaging, transportation management, inventory management and distribution network design. The scope of the operation for each activity was assessed using a three-point scale with three anchors (have outsourced, intend to outsource, and will not outsource). A discrete event that companies choose either to make (in house) or buy (outsourcing) is used as a measure of governance choice. In order to measure the boundary choice more correctly, we coded the status of 'planning to outsource' and 'do not want to outsource' as a 'make' choice for the studied activities.

Asset specificity

Logistics-specific assets were measured using an instrument adapted from Poppo and Zenger (1998) and Roberson and Gatignon (1998). The instruments comprise three-item scales. The Cronbach alpha is 0.69. The scales assess the extent to which the firm commits the investments for each logistics activity. Items are measured using 10-point scales anchored by 'strongly disagree' and 'strongly agree.'

Performance measurement uncertainty

Performance measurement uncertainty was measured using an instrument adapted from Robertson and Gatignon (1998). In our research, two items scales were used. This scale assesses the extent to which the firm evaluates the performance of a logistics service provider. Items are measured using 10-point scales anchored by 'strongly disagree' and 'strongly agree.'

Core closeness

Core business closeness indicates the value of a certain activity to firms. The value refers to efficiency benefits tied to knowledge and skills in executing a certain activity (Barney, 1991; Hafeez *et al.*, 2002). Three-item scales are created and designed with a Cronbach alpha of 0.75 to measure the core closeness of a logistics activity to firms. Items are measured using 10-point scales anchored by 'strongly disagree' and 'strongly agree.'

Supply chain complexity

A new scale was developed to measure supply chain complexity based on literature review (Milgate, 2001; Rao and Young, 1994; Stadtler, 2002; Van der Vorst and Beulens, 2002; Van Goor *et al.*, 2003) and interviews with logistics manager. A list of characteristics (in total 17) that may contribute to the complexity of logistic activities is listed in Table 3.1. Respondents were asked to rate the degree to which the items complicate logistics management in their organisation on a seven-point Likert scale ranging from (1) extremely low to (7) extremely high. As some of the characteristics may be related to each other, we performed a factor analysis. We performed the factor analysis for the total sample as well as for the Dutch and Taiwanese companies separately. After factor analysis six items were dropped due to differences in the distributions of the scores for Taiwan and the Netherlands as well as unacceptable low factor loading (below 0.3). This results in three new meaningful variables:

- *Distribution complexity* which comprises the following items: number of packaging lines, number of clients, delivery frequency, lead-time requirement[3].

---

[3] Period of time between order received and order delivered.

Table 3.1. Results of exploratory factor analysis of supply chain complexity[a,b].

Complexity items	Factor 1	Factor 2	Factor 3
1. Perishability			
2. Number of stock keeping units			
3. Number of product groups			
4. Storage variety		0.700	0.299
5. Number of packaging lines	0.660		
6. Production uncertainty			
7. Demand volumes			0.754
8. Demand uncertainty			0.868
9. Demand fluctuation	0.264		0.839
10. Number of clients	0.711	0.284	0.251
11. Number of international clients			
12. Number of warehouses		0.843	
13. Distribution channel varieties	0.415	0.668	
14. Delivery frequency	0.857		
15. Lead time requirement	0.790	0.254	
16. Distribution size			
17. Distribution uncertainty	0.321	0.723	
Eigenvalue	4.981	1.432	1.094

[a] Factor1=distribution complexity; factor 2=distribution channel complexity; factor 3=demand complexity.

[b] Values less than 0.25 have been omitted; values underlined refer to the significant higher loadings.

- *Distribution channel complexity* which comprises the following items: storage variety[4], number of warehouses, distribution channel variety[5], distribution uncertainty[6].
- *Demand complexity*: demand volumes, demand uncertainty[7] and demand fluctuation[8].

---

[4] Different types of storage requirement.

[5] Inconsistency on route taken by a product as it passes from manufacturer to retailer, for example from manufacturer to distribution center or to retailer shop.

[6] Inconsistency in amount of time and quality for shipments to reach key customers.

[7] Unforeseen inconsistency in demand quantity.

[8] Predictable inconsistency in demand quantity.

Control variables

Geographic location One indicator variable was created to specify whether the focal firm is headquartered in the Netherlands (coded as 1) or Taiwan (coded as 0).

Firm size In this research firm size is measured by number of employees on a national scale.

Sale growth rate changes Respondents are asked to provide sales growth rate changes of their division for the last three years and expected sales growth change level for the next three years. Thus two indicator variables were created to specify whether the sales growth rate is increasing or decreasing, the reference variable is non-change.

3.3.2 Analysis

The objective in this research is to determine the relationship between the firm's transaction and intra-firm's characteristics and levels of outsourcing decision. A great body of literature has suggested a binomial (or binary) choice model to evaluate the relationship between a make-or-buy decision and a set of covariates (Leiblein and Miller, 2003; Robertson and Gatignon, 1998). Thus, we use binary logistic regression to predict a categorical dependent variable and to determine the percent of variance in the dependent variable explained by the independents. The resulting multivariate statistical model takes the following basic form:

> *Buying (outsourcing) decision = $\beta 0$ +$\beta 1$–3 Controls +$\beta 4$ Asset specificity +$\beta 5$ Performance measurement uncertainty + $\beta 6$ Core closeness +$\beta 7$ Distribution complexity +$\beta 8$ Distribution channel complexity +$\beta 9$ Demand complexity + ε (1)*

The likelihood ratio test was used to test the significance of the coefficients. 'Likelihood' is the probability that the observed values of the dependent variables can be predicted from the observed values of the independents. Like any probability, the likelihood varies from 0 to 1. The log likelihood (LL) is its log and varies from 0 to minus infinity. The principle of likelihood ratio test is to compare observed values of dependent variable to predicted values obtained from models with and without the independent variables (Hosmer and Lemeshow, 1989).

3.3.3 Sample description

Table 3.2 shows the respondents' profile. The food processor companies represented in the sample range widely in terms of number of employees. The majority is distributed in the 'fewer than 50 employees' group (27%). Furthermore, these companies range widely in terms of processor type. A large number of respondents is in the 'others' group (29%). Correlation tables are included in Appendix 3.

Table 3.2. Profile of respondents.

Profile		Numbers (N=114)	Percentages
Employees	fewer than 50	31	27%
	50-<100	27	24%
	100-<150	11	10%
	150-<250	28	24%
	larger than 250	17	15%
Plants	1	57	50%
	2	25	22%
	3	9	8%
	4 or larger	23	20%
Sectors	meat	19	17%
	fish	5	4%
	fruit and vegetables	10	9%
	oils and fats	2	2%
	dairy	12	11%
	grain mill	6	5%
	animal feeds	11	10%
	others[a]	33	29%
	beverages[b]	8	7%
	lunch box	5	4%
	prepared meal	2	2%

[a] Others include bread, biscuits, sugar, cocoa, macaroni, coffee, etc.
[b] Beverages include alcoholic, wines, fruit wines, beer, etc.

3.4 Results

Table 3.3 presents the results for the outsourcing decision of four levels of logistic activities. Inventory management is not discussed because there are only a few representative cases in the 'outsourcing' group. Model I is a baseline model that consists of an intercept term and measures of geographic region, firm size and growth rate changes. Model II introduces measures derived from transaction cost theory, resource-based view, and supply chain management theories, i.e. core closeness, asset specificity, performance measurement uncertainty, and supply chain complexity. Likelihood statistics and measures of overall model fit are included at the bottom of the table. For each activity the number of respondents with the make and buy choice are different (see footnote of Table 3.3). The best model is assessed by the improvement of the

Table 3.3. Results of logistics regression analysis of outsourcing decision[a,b,c,d,e].

Levels Activities Independent variables	1st Transportation Model I	Model II	2nd Packaging Model I	Model II	3rd Transportation management Model I	Model II	4th Distribution network design Model I	Model II
Intercept	-1.92 (1.02)+	-3.72 (2.078)+	-3.45 (1.32)**	-5.19 (2.29)*	-1.68 (0.951)	0.063 (1.40)	-1.45 (1.69)	-4.70 (2.70)+
Location (the Netherlands)	1.37 (0.546)*	1.902 (0.816)*	0.287 (0.646)	0.130 (0.821)	0.751 (0.497)	0.526 (0.560)	-0.714 (0.855)	-1.02 (0.912)
Size	0.444 (0.189)*	0.638 (0.240)*	0.432 (0.235)*	0.316 (0.279)	0.151 (0.166)	0.364 (0.203)	-0.283 (0.302)	-0.232 (0.341)
Sales growth (increasing)	0.288 (0.585)	0.673 (0.744)	-0.096 (0.741)	-0.214 (0.852)	0.252 (0.576)	0.335 (0.644)	0.566 (1.136)	0.459 (1.260)
Sales growth (decreasing)	0.273 (0.658)	0.315 (0.826)	-0.866 (0.983)	-1.38 (10.22)	-0.031 (0.654)	0.518 (0.743)	0.881 (1.211)	0.697 (1.312)
Asset specificity		-0.850 (0.212)***		-0.714 (0.240)**		-0.257 (0.140)+		-0.369 (0.258)
Performance measurement uncertainty		0.189 (0.132)		0.040 (0.137)		0.144 (0.106)		0.105 (0.186)
Core closeness		0.597 (0.202)**		0.480 (0.225)*		-0.394 (0.203)+		0.278 (0.251)
Distribution complexity		-0.193 (0.267)		0.785 (0.364)*		-0.005 (0.209)		0.260 (0.312)
Distribution channel complexity		0.040 (0.251)		0.014 (0.278)		0.011 (0.196)		-0.426 (0.315)
Demand complexity		0.119 (0.207)		-0.122 (0.287)		0.001 (0.174)		0.654 (0.345)+

Table 3.3. Continued.

Levels Activities Independent variables	1st Transportation		2nd Packaging		3rd Transportation management		4th Distribution network design	
	Model I	Model II	Model I	Model II	Model I	Model II	Model I	Model II
Log likelihood	123.281	89.846	79.941	60.233	131.769	112.934	58.687	50.015
-2 [L(Intercept)- L(model I)][e]	-17.92 (4)***		-11.12 (4)**		-5.86 (4)		-3.26 (4)*	
-2 [L(model I)- L(model II)]		-66.87 (7)***		-39.42 (7)***		-37.7 (7)***		-17.34 (7)**
Correctly classified (%)	71.4	76.2	85.4	88.3	65.7	71.6	90.9	92.0
Cox & Snell R Square	0.082	0.332	0.053	0.218	0.028	0.192	0.016	0.10

[a] Positive coefficients indicate a greater probability to outsource the activity.

[b] For each variable, the estimated coefficient is given, and standard errors are in parenthesis.

[c] $+P<0.10$; $*P<0.05$; $** P<0.01$; $*** P<0.001$ (2-tailed).

[d] N=114; For Transportation: $N_{outsource}$ =79;$N_{not\ outsource}$ =35 and Log likelihood for null model was 132.239.
For Packaging: $N_{outsource}$ =18;$N_{not\ outsource}$ =96 and Log likelihood for null model was 85.501.
For Transportation management: $N_{outsource}$ =42;$N_{not\ outsource}$ =72 and Log likelihood for null model was 134.701.
For Distribution network design: $N_{outsource}$ =12;$N_{not\ outsource}$ =102 and Log likelihood for null model was 60.318.

[e] Appropriate degrees of freedom are reported in parentheses.

likelihood ratio (-2LL), which reflects the significance of the unexplained variances. For example, the best model for the transportation activity is Model II because adding the five variables significantly improved the model, chi-square $(7, N=114)=66.87$, $P<0.001$. In addition, the correct classification of the outsourcing decisions is improved from 71.4% to 76.2%, and the R^2 statistics have increased in value from 0.082 to 0.332. Given the stability of our results across specifications, we focus on Model II for all activities. Below we present the test results of hypotheses and discussion. Positive coefficients indicate a greater probability of outsourcing.

Asset specificity

The firm's investment in specific logistics asset is a significant predicator of the outsourcing decisions for transportation, packaging and transportation management. As hypothesised, the lower the current investment by firms in transportation ($\beta = -0.850$; $P<0.001$), packaging ($\beta = -0.714$; $P<0.01$), and transportation management activity ($\beta = -0.257$; $P<0.10$), the greater the likelihood that these activities are carried out by logistics companies rather than internally. Asset specificity was found significant in support of theory predictions and confirmed at the 1st, 2nd and 3rd level of activities. The results are consistent with other types of outsourcing activities, such as salesman outsourcing, R&D outsourcing or IT outsourcing (Aubert *et al.*, 1996; Aubert *et al.*, 2004; Hair *et al.*, 1998; Robertson and Gatignon, 1998). This indicates that 'asset specificity' is a good predictor for make-or-buy decisions for logistics activities. In line with the TCA theory's predictions (Williamson, 1998), food companies outsource a certain logistics activity when the logistics facilities or personnel can be redeployed without losing value, thus minimising transaction cost. Hence, proposition 1 is supported in most of the activities studied.

Performance measurement uncertainty

Proposition 2 states that as the performance measurement uncertainty increases, the likelihood of outsourcing decreases because transaction costs associated with negotiating, monitoring and enforcing outsourcing arrangements increase. Data shows that this proposition is not supported at all levels of logistics activities in our study. This is not in line with the theory prediction (Anderson, 1998; Robertson and Gatignon, 1998; Aubert *et al.*, 2004). This implies that the performance measurement uncertainty is not a good predictor for make-or-buy decisions for logistics activities. One explanation for this finding might be that rapidly changing environments may allow LSPs, which specialise in developing a particular technology or process, to capture a higher portion of the economic rents generated by the outsourcing agreement. In this regard in rapidly changing environments, powerful suppliers with specialised skills may be able to exert higher levels of bargaining power over the food processors.

Core closeness

Proposition 3 states that the closer a logistics activity is to the core business, the less likely that a food firm will outsource that activity. The decision to outsource transportation management is negatively related to the core business closeness ($\beta = -0.394$; $P<0.10$), just as hypothesised. Therefore, the 3rd level of outsourcing fits our expectations (Quinn and Hilmer, 1994; Teng *et al.*, 1995; Poppo and Zenger 1998; Leiblein and Miller, 2003).

However, core business closeness is positively associated with an increased incidence of outsourcing for transportation ($\beta = 0.597$; $P<0.01$), and packaging ($\beta = 0.480$; $P<0.05$). These results contradict our expectations. This result reveals that a firm regards transportation and packaging as core business closeness activities but still outsource them. One possible explanation is that within these service markets, LSPs have superior capabilities over food companies to enable efficient service production. Therefore outsourcing these activities is a preferred choice. In such instance, using long-term contracts or cooperating with trustworthy logistics companies would be the best solution.

Supply chain complexity

We identified three dimensions of supply chain complexity, i.e. distribution complexity, distribution channel complexity and demand complexity. Our data show that the decision to outsource packaging and distribution network design indeed depends on the degree of supply chain complexity. In particular, as distribution complexity increases, the likelihood of outsourcing the packaging activity increases as well ($\beta = 0.785$; $P<0.05$). In addition, the probability of outsourcing distribution network design increases as well when a firm is confronted with higher demand complexity ($\beta = 0.654$; $P<0.10$). Thus, proposition 4 is partly supported.

Distribution complexity and packaging Our data show that a firm which owns numerous packaging lines, serves numerous clients under high delivery frequency and strict lead-time requirements, is likely to outsource packaging activities. We notice that some of these items in the distribution complexity are time-related; this may explain the idea behind postponing packaging in the distribution channel for rapid delivery of customised products close to customers (Van Hoek 1999).

Demand complexity and distribution network design Results show that a firm with high demand complexity (demand volume, demand uncertainty and demand fluctuation) is more likely to outsource the distribution network design activity; in other words, decisions to select a road carrier or locate a new factory or warehouse are transferred to a logistics company. Operating under the condition of large demand volume, high demand uncertainty and high demand fluctuation, it might be difficult for a firm to manage its distribution system or production system, thus delivery performance may be damaged. By transferring such problems to a logistics service provider, this service provider can provide flexibilities and effect a great degree of efficiency by exploiting economies of scale among others. Thus in such cases, capacity can be better utilised because the peaks and drops in transport quantities offered by different clients can be counterbalanced, and backhauls are often available to maintain or improve the service level.

Among the complexity items, the 'number of clients', 'lead-time', 'demand volume' are consistent with Rao and Young (1994)'s studies; and the 'delivery frequency', 'demand uncertainty', and 'demand fluctuation' correspond to the findings of Van Damme and Van Amstel (1996). Based on our findings, we assert that the numerousness, variety and interactions complicate logistics planning. As such these complexities require a great effort for food processors in our sample to manage the logistics system (Choi and Krause, 2006). In this regard, outsourcing of a certain logistics activity makes sense when a firm operates under high supply chain complexity, because the operational load to manage such a complex supply chain system increases. In the long term if not well organised, the supply chain responsiveness could be damaged as well. Thus, relying on logistics service providers could be an alternative way of both saving time and avoiding risk.

Control variables

Among the four levels of activities, only the transportation's outsourcing decision is influenced by *geographic region*. It shows that firms in the Netherlands are more likely to outsource transportation than firms in Taiwan ($\beta = 1.90$; $P<0.05$). In addition, the likelihood of outsourcing transportation ($\beta = 0.638$; $P<0.05$) increases if a *firm's size* increases. Furthermore, as expected, none of the activities are found to be related with the change of sales growth rate. Table 3.4 summarises our findings.

3.5 Discussion and conclusions

Regarding our research question: *What decision-making criteria are considered by firms when outsourcing a certain logistics activity?* it can be concluded that the present study delivers a number of interesting results. First, our survey results reveal that determinant factors for logistics outsourcing in the food industry are: asset specificity, core closeness, supply chain complexity (distribution complexity and demand complexity). Second, our results suggest that each level of logistics activities has its own key determinants. Below we summarise our main findings:

- 1st level: The 1st level of logistics activities includes execution activities, such as transportation or warehousing. This level of outsourcing is determined by low asset specificity and high core closeness. The likelihood of outsourcing this level of activities increases, when the transport or storage facilities can be easily redeployed to other uses without losing economic value. However, the positive relationship between core closeness and outsourcing contradicts to our expectation. This can be explained by LSPs having superior capabilities to food processors in this outsourcing market. Therefore, outsourcing

Table 3.4. summary of propositions.

	Propositions
P1: The higher the asset specificity of a specific logistics activity, the less likely that a food processor will outsource this activity than keep it in-house.	Confirmed for level 1, 2, 3; not confirmed for level 4
P2: The higher the performance measuring uncertainty when outsourcing a logistics activity, the less likely that a food processor will outsource this activity than keep it in-house.	Not confirmed for all levels
P3: The closer a logistics activity is to the core business, the less likely that a food processor will outsource this activity than keep it in-house.	Confirmed for level 3; rejected for level 1 and 2; not confirmed for level 4
P4: The higher the supply chain complexity, the more likely a food processor will outsource a logistics activity than keep it in-house.	Distribution complexity: confirmed for level 2 Demand complexity: confirmed for level 4

the 1st level of activities is a preferred choice although food companies regard these activities as central to core business.

- 2nd level: The value-added activities also belong to execution activities. This level of outsourcing is determined by low asset specificity, high core closeness, and high distribution complexity. The likelihood of outsourcing the second level of activities increases, when the packing/labelling equipments or personnel can be easily redeployed to other uses without losing economic value. Besides, food companies tend to outsource this level of activity when they operate in a supply chain setting with high distribution complexity (numerous packaging lines, numerous clients, high delivery frequency and short lead-time) because they can ease or transfer such complexities to LSPs. The positive relationship between core closeness and outsourcing also contradicts our expectations. Similar to the 1st level outsourcing, this can be explained by LSPs having superior capabilities to food processors in the value-added service market.

- 3rd level: The level of activities includes transportation management and inventory management. Our findings show that this outsourcing level is determined by low asset specificity and low core closeness. The likelihood of outsourcing this level of activity increases when the logistics equipments or personnel can be easily redeployed to other uses without losing economic value. In addition, this activity tends to be outsourced when the value of this activity to firm is low, i.e. firm's skills and knowledge are not superior to competitors in enabling an efficient service.

- 4th level: Our results show that this highest level is determined solely by demand complexity. The outsourcing likelihood of distribution network design increases when the firm operates under high demand complexity (demand volume, demand uncertainty and demand fluctuation), because the operational load to manage such a complex supply chain system increases. In the long term if not well organised, the supply chain responsiveness could be damaged as well. Thus, relying on logistics service providers could be an alternative way of both saving time and avoiding risk. However, since the number of companies who have outsourced this level is relatively low, more research is needed in the future to acquire more insight into this level of activities.

A logistics process consists of many different types of activities. Identifying what to outsource is the first step in the outsourcing process. Our findings show that there is no single rule applicable to all logistics activities. Our research suggests evaluating a make-or-buy decision from three different perspectives: transaction cost based, resource based and supply chain logistics. These different perspectives deal with partly overlapping phenomena in complementary ways. For instance, if firms choose to outsource, this may well be due to low transaction costs as well as low efficiency tied to skills and knowledge and complex supply chain settings. When it requires a lot of effort to manage a complicated supply chain, firms may access other resources and complement their own resources with those of other specialists. These result seems to be in line with other studies (Holcomb and Hitt, 2007; Jacobides and Hitt, 2005; Madhok, 2002).

This research provides a detailed look at the logistics outsourcing behaviour of food processors, but it was limited in several ways that might be addressed in future research. The number of firms that had outsourced the 4th level was relatively low. This might be due to the fact that this outsourcing service is still new to the market. We encourage further research to investigate other developed countries, such as USA or Canada and other industries that are expected to have more 4th levels of outsourcing cases.

Chapter 4. The impact of the level of logistics outsourcing on service performance[9]

4.1 Introduction

This chapter aims to answer the following research question.

Research question 3

RQ3 What is the impact of logistics outsourcing on service performance?

This study seeks to advance our understanding of the relationship between the outsourcing decision, the outsourcing level and a firm's logistics service performance. We sought answers to the following questions: *Does logistics outsourcing enhance logistics service performance? Does outsourcing different types of logistics activities have different service outcomes? Do greater levels of outsourcing result in better performance?*

To achieve our research objective, we examined not only the general effect of outsourcing on service performance but also how the supply chain logistics environment moderates the relationship between outsourcing and service performance. Our article is organised as follows: Section 4.2 presents the literature review and Section 4.3 develops hypotheses related to the direct effect of logistics outsourcing and the moderating effect of supply chain complexity on service performance. Section 4.4 presents the research design, providing details on data collection and the constructs applied in this research. The results of our investigations on direct and moderating effects are presented in Section 4.5. Finally, Section 4.6 discusses our findings, sketches in the research limitations and makes suggestions for further research.

4.2 Literature review

Various theoretical perspectives on potential benefits explain why firms engage in outsourcing (e.g., Barney, 1991; Poppo and Zenger, 1998; Williamson, 1975): for example, focus on core business, risk reduction, cost-savings, time-savings, etc. Table 4.1 provides a brief overview of some empirical studies on actual benefits. In this section we discuss some previous studies on outsourcing effects with a focus on manufacturing firms. Two categories are discussed: core business outsourcing and non-core business outsourcing.

---

[9] This chapter is based on an article submitted to an international scientific journal: Hsiao, H.I., R.G.M. Kemp, J.G.A.J. van der Vorst, S.W.F. Omta, The impact of level of logistics outsourcing on service performance in the food processing industry.

Table 4.1. Literature review: outsourcing and performance in the manufacturing industry[a].

Activities	Sources	Performance metrics	Results of main effect	Is there any moderating effect discussed?
Core business:				
Manufacturing	Dabhilkar and Bengtsson (2008)	Quality, speed, dependability, flexibility, cost (at plant level)	Positive effect: volume flexibility; Negative effect: quality, speed and on-time delivery	No
Manufacturing	Dankbaar (2007)	Innovation capability	Negative effect	No
Manufacturing	Jiang et al. (2007)	Market value	Positive effect	No
Non-core business:				
Human resource	Gilley et al. (2004)	Financial performance; Innovation performance; Stakeholder performance	Positive effect: innovation performance	Yes (firm size)
Logistics	Power et al. (2006)[a]	Customer satisfaction, inventory control, capacity management, productivity, service quality, flexibility, sales growth, net profit, cycle times, cash flow, general cost management, backlog management and transportation cost management	Positive effect: cost management and flexibility	No
Logistics	Larson and Kulchitsky (1999)[a]	Customer service, cost reduction	No direct effect	No
Not specified				
Not specified	Gilley and Rasheed (2000)	Financial, innovation, stakeholders	No direct effect	Yes (corporate strategy and environmental dynamism)
Not specified	Jiang et al. (2006)[a]	Cost efficiency, productivity, profitability	Positive effect: cost efficiency	No
Not specified	Salimath et al. (2008)[a]	Profitability, sales revenue, net profit, growth in profits and in sales revenue	Positive effect on all	Yes (organisational configuration: age, size, innovation, ownership)

[a] Studies that cover wide ranges of industries.

4.2.1 Core business outsourcing

Current studies have related positive effects of manufacturing outsourcing to production volume flexibility or market value, but negative effects to innovation capabilities, quality, speed, and on-time delivery. For example, Dabhilkar and Bengtsson (2008) found positive direct effects of outsourcing on volume flexibility. The focal firm can improve its responsiveness to variability in demand by outsourcing peak demand to suppliers. According to Jiang *et al.* (2007) core business-related outsourcing is positively related to outsourcing firms' market value. It demonstrates a positive signal to the stock market. They mention that firms, recognising that they cannot be world class in every activity and function involved in producing their products, are moving toward business strategies based on 'core competencies' that help maintain their competitive advantage in serving customers.

However, core business outsourcing tends to be negatively related to innovation capabilities. Dankbaar (2007) investigates the relationship between manufacturing outsourcing and innovation. He indicates that the long-term impact may well be a loss of innovative capabilities on the part of the outsourcing company because product development follows manufacturing. Furthermore, he explains that if manufacturing is done in another company, access to manufacturing knowledge by development people will tend to become more difficult. This may result in less producible products. In addition, Dabhilkar and Bengtsson (2008) also found that manufacturing outsourcing might have negative effects on quality, speed and on-time delivery.

4.2.2 Non-core business outsourcing

The non-core businesses in manufacturing industry include IT, human resource management, accounting or other financial services, and logistics or transportation. In general, they are not obviously relevant to manufacturing firms' core business. This section discusses non-core business outsourcing and also includes the studies that include both core and non-core outsourcing (Table 4.1). Our reading of this literature brings us to the following conclusions:

First, outsourcing of non-core business has positive impacts on innovativeness (Gilley *et al.*, 2004), cost efficiency (Jiang *et al.*, 2006; Power *et al.*, 2006), profitability (Salimath *et al.*, 2008), and logistical flexibility (Power *et al.*, 2006). Some of these benefits are rarely mentioned in relation to core business outsourcing, as we discussed earlier. For example, Gilley *et al.* (2004) found that firm innovativeness is related to higher levels of HR outsourcing. Both training and payroll outsourcing were found to be significant predictors of innovation performance, for instance R&D outlays, process innovations, and product innovations. Although these results do not warrant an inference of a cause–effect relationship, they seem to indicate that firms may potentially achieve higher levels of focus on those activities that drive innovation and other forms of competitive advantage by entrusting training and payroll activities to outside specialists. Jiang *et al.* (2006) provide empirical evidence of the difference between outsourcing firms' performance and that of their non-outsourcing competitors. Outsourcing firms have advantages in cost efficiency over their counterparts who do not outsource any activities at the same time. Jiang *et al.* (2006) explain that outsourcing arrangements that transfer outsourcing firms' assets to a vendor can convert fixed amortisation and operating expenses to variable usage charges. On the application side, outsourcing can reduce the commitment to fixed-cost, full-time HR expenses and other overhead costs through contracts that provide development skills on an as-needed basis. As a result, outsourcing can improve firms' cost efficiency.

Second, there is an increasing focus on the moderating effect of organisation configuration and environmental dynamism on outsourcing performance (Gilley *et al.*, 2004; Gilley and Rasheed, 2000; Salimath *et al.*, 2008). For example, Salimath *et al.* (2008) found that outsourcing not only has a positive effect on financial performance, but also the outsourcing-performance relationship is moderated by different configurations, such as size or age. For instance, their findings suggest that outsourcing tactics result in the greatest benefit to the large firms due to their ability to manage resource dependency relationships. Besides size advantages, larger firms can negotiate better terms in the outsourcing contract, through volume discounts. If left dissatisfied, larger clients can influence suppliers more than smaller firms can by taking their business elsewhere.

4.3 Theoretical framework

In this section, we extend previous arguments to logistics outsourcing and build our theoretical framework. Figure 4.1 shows the relationships between the direct effect of outsourcing on service performance, and the moderating effects of supply chain complexity on outsourcing-performance relationships. A number of authors mention that the logistics outsourcing decision is especially related to supply chain complexity (Hsiao *et al.*, 2006; Milgate, 2001; Rao and Young, 1994), thus we include this factor as a moderator in our research model. We will discuss this factor in more detail later.

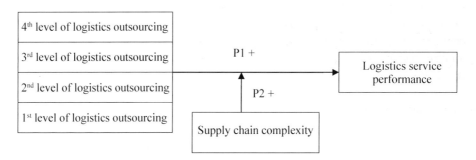

Figure 4.1. Research model: main effects (P1) and moderating effects (P2).

4.3.1 Definitions

Levels of logistics outsourcing

Logistics is a process of planning, implementing and controlling the efficient, cost-effective flow and storage of raw materials, in-process inventory, finished goods, and related information from point of origin to point of consumption for the purpose of conforming to customer requirements (Van Goor *et al.*, 2003). Logistics activities which can be outsourced range from execution activities, such as transportation, to planning activities, such as transportation planning (Dapiran *et al.*, 1996; Hong *et al.*, 2004; Millen *et al.*, 1997; Razzaque and Sheng, 1998; Sahay and Mohan, 2006; Sohail *et al.*, 2006; Wilding and Juriado, 2004). In our research, the outsourceable logistics activities are divided into four levels (Hsiao *et al.*, 2006):

Execution activities
- *Level 1*: Activities include transportation and warehousing. At this lowest level, contractual relationships between LSPs and their clients are often short term.
- *Level 2*: Activities include value-added activities, which refers to tasks normally performed by processors but now being moved into distribution as part of final processing. In the food manufacturing industry, these tasks include mixing flavours, packaging or labelling. The contractual relationships between LSPs and their clients are often limited to one year or less.

Planning activities
- *Level 3*: This refers to the outsourcing of logistics planning and control activities, such as inventory management and transportation management. The LSPs offer customised logistics solutions and their skills are complementary to that of their clients.
- *Level 4* (total outsourcing): This refers to outsourcing the distribution network design. At this strategic planning and control level, decisions are made concerning supply chain restructuring, for example, changes in the warehouse structure, reassignment of tasks between tiers, redistribution of inventory between tiers, changes in transportation network, mode, consolidation points, reassignment of roles and responsibilities among chain entities. When activities at this level are outsourced, the LSPs take care of the logistics network design and orchestrate the logistics flow of the network (Van der Vorst *et al.*, 2007).

Definition: logistics service

Logistics creates value by accommodating customer's delivery requirements in a cost effective manner (Stank *et al.*, 2003). The system to measure logistics performance may cover the following areas: internal performance within the units (for example, materials management, production and distribution), external performance within the units, external performance of the entire company towards the customer, supplier performance towards the company, and the relationship between the logistics performance and the performance of the entire company (Andersson *et al.*, 1989). Logistics service belongs to the 'external performance of the entire company towards the customer.'

Logistics service performance assesses a provider's ability to consistently deliver requested products within the requested delivery time frame at an acceptable cost (Bowersox and Closs, 1996). Service performance is a very wide term and varies from one company to the next (Kisperska-Moron, 2005; Stank *et al.*, 2003). Moreover, suppliers and customers often hold differing views on this concept. In this chapter, we use the following indicators of logistics service to measure the impact of outsourcing: reliability, flexibility, and lead-time (Kisperska-Moron, 2005; Stank *et al.*, 2003; Wilding and Juriado, 2004). In the past, authors often used these indicators for logistics service measurement.

4.3.2 Direct effect of logistics outsourcing

By outsourcing non-core business, firms may enhance their performance by focusing on core activities that allow them to become more innovative in their core business (Arnold, 2000; Gilley *et al.*, 2004; Jiang *et al.*, 2006; Prahalad and Hamel, 1990; Quinn and Hilmer, 1994). We extend these arguments to logistics activities. Logistics is often regarded as non-core business in the food manufacturing industry; therefore we argue that a food company engages in outsourcing in order to derive benefits.

Many studies have shown that logistics outsourcing has a positive impact on logistics performance, particularly on cost (Capgemini, 2007; Norek and Pohlen, 2001; Power *et al.*, 2006; Van Damme and Van Amstel, 1996). Cost reduction from logistics outsourcing comes mainly from better utilisation of capacity and better capital allocation. Capacity can be better utilized by the service provider because the peaks and drops in transport quantities offered by various clients can be counterbalanced, and because backhauls are often available. Thus the service provider can effect a great degree of efficiency, by exploiting economies of scale, among other things. In addition, manufacturers can also better allocate their capital by, for example, refraining from investing in storage or trucks for the purpose of distribution capacity, which may reduce risk.

Although few studies related to logistics outsourcing have focused on service performance, we argue that service performance is an important area to focus on because logistics outsourcing is often expected to influence service performance. For example, it could provide greater flexibility in adapting to changes in the market (Power *et al.*, 2006). When demand surges beyond a firm's own capability, a third party may be called in to help meet the increased demand (Razzaque and Sheng, 1998). In addition, lead-time reduction could be another potential benefit of logistics alliances. Long lead-time is often a problem and requires large inventories in transit and at the sales subsidiary (Bhatnagar and Viswanathan, 2000; Halldorsson and Skjott-Larsen, 2004). Through logistics outsourcing, LSPs can help clients to reduce lead time by means of several restructuring strategies, including faster modes of transportation, more direct transport or eliminating local inventory stocking points.

Overall, outsourcing seems not only to show positive benefits for cost reduction, but also in service performance. Thus, we suggest that by outsourcing logistics activities, firms can achieve better service performance relative to firms that do not outsource those activities. We formulate the following proposition:

> *Proposition 1 Logistics outsourcing has a positive effect on a firm's logistics service performance.*

4.3.3 Moderating effect: supply chain complexity

Contingency theory is about analysis of the relationship between organisations and their environment (McAuley *et al.*, 2007; Perrow, 1967). The central theme of contingency theory is that all components of an organisation must 'fit' well with each other otherwise the organisation will not perform optimally (Perrow, 1967). A contingency theory differs from other theories in the specific form of its propositions (Drazin and Ven, 1985). In a congruent proposition a simple unconditional association is hypothesised to exist among variables in the model. A contingent proposition is more complex, because a conditional association of two or more independent variables with a dependent outcome is hypothesised and directly subjected to an empirical test. Different external conditions may require different organisation characteristics and behaviour patterns within the effective organisation (Lawrence and Lorsch, 1967). Therefore, to investigate the relationships between outsourcing decision, performance and supply chain complexity, we use the concept of contingency theory.

Supply chain complexity, from a logistical view, refers to the level and type of interactions present in the upstream and downstream of logistics flow (Milgate, 2001). It is determined by its extent, variety (differentiated, inconsistent) and variability (predictability of changes) (Germain *et al.*, 2008; Hsiao *et al.*, 2008). For example, supply chain complexity is high when the numbers of suppliers or customers increase and regions for delivery are inconsistent. There are four main sources of logistics complexity that can plague a firm in a supply chain: (1) upstream sources such as number of suppliers, volume, time and material quality and uncertainty; (2) internal sources related to manufacturing such as volume of throughput, output, product quality and uncertainty; (3) internal sources related to distribution such as volume of distribution, and uncertainty; (4) downstream sources related to customers, such as changes in orders and demand uncertainty (Ball, 2007; Londe and Cooper, 1998; Stadtler, 2002; Van der Vorst and Beulens, 2002; Wanke and Zinn, 2004).

Few studies have investigated supply chain complexity as a moderator to outsourcing service performance relationships. Only one study has addressed the contingent effect of environment on the relationship between outsourcing and cost performance. Gilley and Rasheed (2000) propose a negative effect of environmental uncertainty on outsourcing and cost performance because transaction costs associated with negotiating, monitoring, and enforcing outsourcing arrangements increase in more dynamic environments. In addition, powerful suppliers with specialised skills may be able to exert higher levels of bargaining power over outsourcing firms in rapidly changing environments. In the following discussion we consider the effects of environmental uncertainty on the behaviour of decision-makers. We go on to illustrate the relationships between supply chain complexity, outsourcing and performance.

Decision-makers working in highly uncertain environments tend to encounter great burdens of information processing. As a result, these individuals are likely to experience high levels of stress and anxiety (Waldman *et al.*, 2001). In general, an uncertain environment may influence decision-making behaviour in two ways: it speeds up the decision-making process and stimulates rational choices. First, from an information perspective, when environmental uncertainty is high organisations need to process more information in order to make decisions. The decision-makers tend to engage in fast-decision behaviours to cope with the anxiety and build confidence (Eisenhardt, 1989; Judge and Miller, 1991). Rapid strategic decision-making helps executives learn fast and capitalise quickly on market opportunities.

Second, decision-makers also tend to make rational choices in uncertain environments. Rationality is the use of information for the purpose of selecting a sensible alternative in the pursuit of one's goal (Dean and Sharfman, 1993). A study by Oh and Rhee (2008) looks at the relationship between technology uncertainty and carmaker decision-making behaviour. They contend that when technology uncertainty grows, the carmaker's cost-overrun risk rises; and consequently, the carmaker gives more weight to the supplier's cost-reduction capability in selecting a partner. Thus, the use of rational processes in dynamic environments assists managers in identifying relevant opportunities, and devising successful responses. In more stable environments, managers use existing information and mental models to formulate effective decisions (Hough and White, 2003).

Deriving our argument from previous discussions on outsourcing, we suggest that supply chain complexity (including uncertainties) may have an impact on the relationship between outsourcing and service performance. Under uncertain complex environments, food processors may tend to use a rational process when selecting a service provider, for example, by giving more weight to LSP service capability. In uncertain environments, LSPs might specialise in developing or implementing a particular technology or process that provides innovative solutions for food processors. Therefore, we propose that the benefits of outsourcing increases with increasing levels of supply chain complexity. Conversely, in low complexity environments, the benefits of outsourcing decline.

Proposition 2 Firms operating in a supply chain setting with high logistical complexity gain greater logistics service performance benefits from logistics outsourcing.

4.4 Data

This research used a combination of Dutch (NL) and Taiwanese (TW) data for several reasons. First, the purpose of this research is to study outsourcing impact on food processors' performance. Dutch food processing is famous worldwide, and agriculture is also an important industry for Taiwan. Second, both countries have geopolitically limited access to natural resources and land. As a response to these limitations, outsourcing can be an effective strategy for processing industries in both countries.

We mailed a total of 890 questionnaire to members of the Dutch Chamber of Commerce (www.ksv.nl) and the Taiwanese Industry & Technology Intelligence Service (www.itis.org. tw) (NL: 385; TW: 505). Our procedures for survey design included a literature review and several interviews with logistics professionals. A draft of the survey instrument was completed by a small group of logisticians. The survey administration entailed two waves of mailings, with all non-respondents to the first wave receiving a second-wave replacement questionnaire. A total of 66 surveys were returned as undeliverable, or from recipients disqualifying themselves as respondents (NL: 57; TW: 9). In total, 138 usable responses were received (NL: 76; TW: 62), of which 24 had missing data (NL: 7; TW: 17) and were judged unusable, thus yielding an effective sample size of 114 (NL: 69; TW: 45) for a response rate of 114/800=15% (NL: 21%; TW: 9%).

4.4.1 Measures

Outsourcing decision

The outsourcing effect is examined by the differences between the outsourcing firm's performance and their non-outsourcing competitors. Respondents were first asked to identify the major product group in terms of turnover and were then asked to describe the current outsourcing status of four logistics activities. The four logistics activities studied are transportation (level 1), packaging (level 2), transportation management (level 3), and distribution network design (level 4). The survey asked respondents to describe their current practice of logistics outsourcing in their main product group given three choices. The choices included: 'we have already outsourced this activity', 'we intend to outsource this activity' and 'we don't intend to outsource this activity'. In order to measure the outsourcing choice more correctly, we coded the status of 'intend to outsource' and 'don't intend to outsource' as current 'in-house' choices.

Supply chain complexity

The literature and logisticians suggest 11 logistics complexity items which might complicate a food manufacture's logistics process (Ball, 2007; Hsiao *et al.*, 2008; Londe and Cooper, 1998; Stadtler, 2002; Van der Vorst and Beulens, 2002; Wanke and Zinn, 2004). Table 4.2 presents

Table 4.2. Results of exploratory factor analysis[a,b].

Factor name	Items	Factor 1	Factor 2	Factor 3
A. Supply chain complexity items: *Rate the extent to which the following items complicate your logistics planning. Likert 7-point scales ranging from 'strongly agree' to 'strongly disagree'.*				
Distribution complexity	Number of packaging lines	0.660		
(α = 0.82)	Number of clients	0.711	0.284	0.251
	Delivery frequency	0.857		
	Lead times	0.790	0.254	
Distribution channel	Storage variety		0.700	0.299
complexity	Number of warehouses		0.843	
(α = 0.79)	Distribution channel varieties	0.415	0.668	
	Distribution uncertainty	0.321	0.723	
Demand complexity	Demand volume			0.754
(α = 0.83)	Demand uncertainty			0.868
	Demand fluctuation	0.264		0.839
	Eigenvalue	4.981	1.432	1.094
B. Performance items: *Compared to your competitors, rate the performance of your major product group. Likert 7-point scales ranging from 'strongly agree' to 'strongly disagree'.*				
Service	We always meet the promised delivery time (reliability 1)	0.807		
(α = 0.81)	We always deliver the ordered quantity (reliability 2)	0.857		
	We quickly respond to the needs of our key customers (flexibility)	0.759		
	We offer a shorter lead time (lead times)	0.783		
	Eigenvalue	2.576		

[a] Values less than 0.25 have been omitted; values underlined refer to the significant higher loadings.
[b] This factor analysis used principle axis extraction techniques and an oblique rotation.

these items. Value less than 0.24 have been omitted (Gilley and Rasheed, 2000). Respondents were asked to rate the degree to which the item complicates logistics management in their product group on a seven-point Likert scale ranging from (1) extremely low to (7) extremely high. We performed factor analysis because some items may be related to others. Factor analysis resulted in three new meaningful variables: distribution complexity, distribution channel complexity, and demand complexity.

- *Distribution complexity* comprises four items: number of packaging lines, number of clients, delivery frequency, and lead times.[10] It describes basic supply chain settings operated by a food processor and routine distribution characteristics (information) which are predictable and often known in advance.
- *Distribution channel complexity* comprises four items: storage variety,[11] number of warehouses, distribution channel variety,[12] and distribution uncertainty.[13] It describes the characteristics of the distribution channel between the factory, warehouse and customers.
- *Demand complexity* comprises three items: demand volumes, demand uncertainty[14] and demand fluctuation[15]. It describes customers' demand characteristics.

Logistics service performance

Logistics service performance assesses a firm's ability to deliver requested products within the requested delivery time frame at an acceptable cost (Stank *et al.*, 2003). Respondents were asked to indicate how their firm's lead-time, reliability, flexibility in their major food product compared with their competitors (Beamon, 1999; Bhatnagar and Viswanathan, 2000; Halldorsson and Skjott-Larsen, 2004; Power *et al.*, 2006), on a Likert scale ranging from strongly disagree (1) to strongly agree (7). Table 4.2 shows that exploratory factor analysis results in one construct. This service construct consists of the following subsets: (1) Flexibility: in responding to marketplace changes, the ability of firms to gain or maintain competitive advantage. It includes volume flexibility and time flexibility, and the willingness to help customers and provide prompt service. (2) Reliability: the ability to perform the promised service dependably and accurately, that is, to deliver the correct product to the correct place at the correct time in the correct condition. (3) Lead time: the speed at which firms provide products to the customer.

Control variables

Three variables are considered control variables: geographic location, size of firm and chilled requirement. *Geographic location* was used to specify whether the focal firm is headquartered in the Netherlands (coded as 1) or Taiwan (coded as 0). *Firm size* was measured by number of employees on a national scale on a Likert scale, ranging from (1) smaller than 50 to (4) larger than 250. *Chilled requirement* was used to specify whether the product group is stored under chilled conditions (coded as 1) or not required (coded as 0).

---

[10] Period of time between order received and order delivered.

[11] Different types of storage requirement.

[12] Inconsistency on route taken by a product as it passes from manufacturer to retailer, for example from manufacturer to distribution center or to retailer shop.

[13] Unforeseen inconsistency in amount of time and quality for shipments to reach key customers.

[14] Unforeseen inconsistency in demand quantity.

[15] Predictable inconsistency in demand quantity.

4.4.2 Analysis and sample description

The objective of this study is to discuss the direct effect of logistics outsourcing on logistics service performance and the moderating effects of supply chain complexity. To test the moderator role of supply chain complexity between outsourcing decisions and service performance, we used multiple hierarchical regression analysis (Jaccard and Turrisi, 2003).

The survey also included several questions about the respondents and their organisations (Table 4.3). The majority of organisations hire 'fewer than 50 employees' (27%). On average the firms owned only one plant. The majority of respondents produce foods categorised as 'others' (28.9%), followed by meat (17%) and dairy products (11%). The table also shows participants and percentages at different levels of activities. The most popular outsourced activity is transportation (level 1), followed by transportation management (level 3), packaging (level 2) and distribution network design (level 4). The number of respondents who outsourced transportation is 79 out of 114. The number of respondents who outsourced transportation management activity is 42. The number of respondents who outsourced the other two activities is relatively low: 18 respondents for packaging and 12 for distribution network design.

4.5 Results

Mean standard deviations and correlations are presented in Table 4.4. To reduce the problem of multicollinearity between predictors and the interaction terms containing these predictors, we employed the mean centering technique, which is the raw score minus the mean of the independent variables (Jaccard, 2003).

Table 4.5 presents the test results of the direct effect of logistics outsourcing on service performance and the moderating effect of the three complexities constructs. Model selection depended on the significance of the F test and F change. Model 1 presents the results of direct effect tests of level 1 outsourcing. The moderating effect is not discussed because the F change (ΔF) shows no significance with the added interaction terms. The direct effects and moderating effects of level 2 and level 3 outsourcing are also not presented as well because these test results are similar to level 1. Model 2 presents the direct effects of level 4 outsourcing, while Model 3 shows the moderating effects of the three complexities constructs. In the regressions, we also added the cost strategy as a control variable. However, this variable was not significant at all levels of activities and the model was not better if included. Thus this variable was not discussed in the results. The following presents our main findings in detail.

4.5.1 Control variables

Table 4.5 shows that the chilled requirement was related to service performance in all activities (model 1: $\beta = 0.582$; $P<0.001$, model 2: $\beta = 0.564$; $P<0.01$ and model 3: $\beta = 0.528$; $P<0.01$).

Table 4.3. Profile of respondents: numbers and percentages.

Profile	Total	Outsourced group			
		1st level	2nd level	3rd level	4th level
	(N=114)	(N=79)	(N=18)	(N=42)	(N=12)
Location					
Netherlands	69 (60%)	52 (66%)	9 (50%)	28 (67%)	7 (58%)
Taiwan	45 (40%)	27 (34%)	9 (50%)	14 (33%)	5 (42%)
Employees					
<50	31 (27%)	21 (27%)	1 (6%)	11 (26%)	3 (25%)
50-<100	27 (24%)	16 (20%)	3 (17%)	10 (24%)	4 (33%)
100-<150	11 (10%)	8 (10%)	2 (11%)	3 (7%)	1 (8%)
150-<250	28 (25%)	20 (25%)	8 (44%)	11 (26%)	2 (17%)
250+	17 (15%)	14 (18%)	4 (22%)	7 (17%)	2 (17%)
Plants					
1	57 (50%)	37 (47%)	6 (33%)	23 (55%)	6 (50%)
2	25 (22%)	17 (21%)	5 (28%)	9 (21%)	-
3	9 (8%)	7 (9%)	2 (11%)	3 (7%)	2 (17%)
4 or larger	23 (20%)	18 (23%)	5 (28%)	7 (17%)	4 (33%)
Chilled requirement					
No	65 (57%)	42 (53%)	12 (67%)	22 (52%)	9 (75%)
Yes	46 (40%)	34 (43%)	5 (28%)	18 (43%)	2 (17%)
Sectors					
meat	19 (17%)	13 (17%)	1 (6%)	5 (12%)	2 (17%)
fish	5 (4%)	4 (5%)	1 (6%)	1 (2%)	-
fruit and vegetables	10 (9%)	7 (9%)	1 (6%)	5 (12%)	1 (8%)
oils and fats	2 (2%)	1 (1%)	-	-	-
dairy	12 (11%)	11 (14%)	2 (11%)	6 (14%)	-
grain mill	6 (5%)	6 (8%)	2 (11%)	3 (7%)	1 (8%)
animal feeds	11 (10%)	4 (5%)	1 (6%)	-	-
others	33 (29%)	23 (29%)	5 (28%)	15 (36%)	6 (50%)
beverages	8 (7%)	7 (9%)	4 (22%)	4 (10%)	1 (8%)
lunch box	5 (4%)	1 (1%)	-	1 (2%)	-
prepared meal	2 (2%)	1 (1%)	-	1 (2%)	1 (8%)

Table 4.4. Descriptive statistics and correlations.

	Mean	S.D.	1	2	3	4	5	6	7	8	9	10
1. Location (the Netherlands)	1.39	0.491										
2. Size	3.7632	1.45920	0.515**									
3. Chilled requirement	0.41	0.495	-0.084	-0.060								
4. Outsourcing decision (level 1)	0.69	0.463	-0.163	0.114	0.099							
5. Outsourcing decision (level 2)	0.16	0.366	0.093	0.253**	-0.104	0.288**						
6. Outsourcing decision (level 3)	0.37	0.485	-0.096	0.037	0.054	0.508**	0.318**					
7. Outsourcing decision (level 4)	0.11	0.308	0.015	-0.023	-0.157	0.228*	0.322**	0.331**				
8. Distribution complexity	4.1091	1.44664	-0.116	0.034	-0.139	-0.063	0.194*	0.051	0.004			
9. Distribution channel complexity	3.4299	1.54249	0.036	0.204*	0.005	0.111	0.185*	0.055	-0.012	0.594**		
10. Demand complexity	4.5929	1.56013	-0.028	0.108	-0.146	0.128	0.099	0.107	0.170	0.432**	0.436**	
11. Service performance	5.4912	0.93655	-0.238*	-0.240*	0.298**	-0.124	-0.099	0.022	-0.073	-0.003	-0.180	-0.172

Table 4.5. Results of hierarchical regression analysis[a,b].

	Logistics service performance		
	Model 1[c,d] **Level 1** **outsourcing**	**Model 2** **Level 4** **outsourcing**	**Model 3** **Level 4** **outsourcing**
Intercept	6.27 (0.450)***	6.07 (.426)***	5.97 (0.424)***
Control variables			
Location (The Netherlands)	-0.327 (0.208)	-0.248 (0.202)	-0.308 (0.200)
Size	-0.053 (0.069)	-0.073 (0.069)	-0.042 (0.069)
Chilled	0.582 (0.172)***	0.564 (0.175)**	0.528 (0.173)**
Moderators			
Distribution complexity (V1)	0.111 (0.077)	0.135 (0.076)*	0.147 (0.075)*
Distribution channel complexity (V2)	-0.123 (0.071)*	-0.137 (0.071)*	-0.123 (0.071)*
Demand complexity (V3)	-0.050 (0.062)	-0.062 (0.063)	-0.066 (0.063)
Main effect			
Outsourcing decision: level 1 (V4)	-0.271 (0.190)		
Outsourcing decision: level 4 (V5)		-0.001 (0.286)	-0.335 (0.309)
Interaction terms			
V1*V5			0.263 (0.237)
V2*V5			-0.163 (0.271)
V3*V5			0.480 (0.195)**
R^2	0.22	0.20	0.26
adj. R^2	0.16	0.15	0.19
Model F	4.02***	3.73***	3.54***
ΔF			2.67*

[a] For each variable, the estimated coefficient is given, and the standard errors are in parentheses.
[b] +$P<0.10$; *$P<0.05$; **$P<0.01$;***$P<0.001$ (one-tailed).
[c] For level 1 outsourcing decisions (model 1), the interaction effects are not included in the table because the ΔF shows no significance difference when the interaction terms are added.
[d] For level 2 and 3 outsourcing decisions, both direct and interaction effects are not included in the table because the results are comparable to those in level 1.

Chilled foods were found to have higher service performance than non-chilled foods, which is reasonable since chilled products are more perishable in nature and thus have greater need for speed, flexibility and reliability than non-chilled products. In addition, firms in Taiwan tend to have higher service performance than firms in the Netherlands; and the smaller firms tend to have higher service performance than in the Netherlands. But the statistics of location and firm size showed no significant influence on service performances in all activities.

4.5.2 Direct effect

Proposition 1 was tested to examine the extent to which an outsourcing decision influences a firm's service performance. In each activity, location, size and chilled requirements were used as control variables. The R^2 is the percent of variance in the dependent explained uniquely or jointly by the independents. Model 1 and model 2 present the test results for level 1 and level 4 activities. Linear combinations of the predictors, adjusted for the number of independent variables, explained 16% of the variance in performance for level 1 outsourcing; and 15% of the variance in performance for level 4 outsourcing. In these activities, the majority of the variance was explained by supply chain complexities. For level 1 outsourcing this was distribution channel complexity and for level 4 outsourcing this was both distribution complexity and distribution channel complexity. However, the table shows that none of the outsourcing decisions significantly influenced service performance. Thus, proposition 1 is not supported, indicating that in our sample there is no direct effect of outsourcing decisions on service performance. The findings are in line with some other market surveys. Capgemini (2007) reports that some service users have chronic problems with LSPs. Often clustered at the top of the problem list is 'service level improvement not achieved'. In addition, a study by Wilding *et al.* (2004) also indicates that most companies report no change to service levels due to outsourcing.

Model 1 and model 2 both show that distribution complexity is positively related to service performance, but distribution channel complexity and demand complexity are negatively related to service performance. This can be explained by the unpredictable environment (the distribution uncertainty item in distribution channel complexity; and demand uncertainty item in demand complexity), which could jeopardise logistics service performance (Guimaraes *et al.*, 1999; Mapes *et al.*, 2000; Milgate, 2001). For example, unplanned demand changes may delay production and increase average throughput times. Production-time variability of the individual manufacturing stages will complicate the task of coordinating manufacturing stages. This will cause longer than expected processing times and also increase the average manufacturing and distribution throughput time, thus influencing delivery speed and flexibility.

4.5.3 Moderating effect

Proposition 2 proposed that the effect of logistics outsourcing on logistics performance was dependent on the degree of supply chain complexity. The complexity was proposed to enhance the positive effects of outsourcing on service performance. We chose model 3 because the F change (ΔF) shows significance with the added interaction terms. Demand complexity was found to interact positively with level 4 outsourcing on service performance ($\beta = 0.48, P<0.01$). The positive coefficient of the interaction terms demonstrates that with increasing levels of demand complexity, outsourcing of level 4 activities has a positive association with service performance. Figure 4.2 illustrates that service performance for the outsourcing decision differs between firms according to high and low demand complexity. Thus, proposition 2 is supported at level 4 outsourcing. This can be explained by the following reasons. Under conditions of high demand uncertainty, food processors are likely to put more emphasis and spend a greater amount of time and resources on environmental scanning (Bstieler, 2005). For example, selecting an LSP with high service capabilities (Oh and Rhee, 2008). In addition, LSPs may also tend to specialise in developing a particular technology or process to provide innovative solutions for food processors (Gilley *et al.*, 2004). Therefore, the benefits of outsourcing increase when the degree of demand complexity increases.

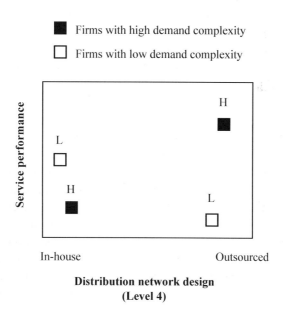

Figure 4.2. Interaction graph of outsourcing decision showing complexity of service performance.

4.6 Discussion and conclusions

The goal of this study was to understand how logistics outsourcing decisions affect logistics service performance. We examined four levels of logistics activities and also assessed the moderating effect of supply chain complexity. This study delivers a number of interesting results. First, all the logistics activities studied had no direct outsourcing effect on service performance. This may be roughly in line with some other market surveys, which often indicate that the main problem with LSPs is a lack of service level improvement (Capgemini, 2007; Wilding and Juriado, 2004). Second, unpredictable and complicated environments decrease service performance levels. Distribution channel complexity (storage variety, number of warehouses, distribution channel variety, and distribution uncertainty) and demand complexity (high production volumes, high demand uncertainty and high demand fluctuation) lower service performance, but distribution complexity (number of packaging lines, number of clients, delivery frequency and lead-time requirement) improves service performance. This may possibly be explained by unplanned changes (distribution uncertainty and demand uncertainty) that often delay production and distribution schedules and increase the average throughput times (Mapes *et al.*, 2000). Thus this decreases service performance. Third, we found that the relationships between outsourcing and service performance are moderated by demand complexity at level 4. This indicates that service performance of outsourcing at level 4 increases with an increasing degree of demand complexity. The possible interpretation is that in a highly unpredictable and complicated environment, food processors tend to select LSPs with high service capabilities to cope with such uncertain situations. In addition, LSPs might also tend to specialise in developing a particular technology or process to provide innovative solutions for food processors. Thus, an unpredictable environment increases the benefits of level 4 outsourcing.

Our findings reveal that only total outsourcing shows service benefits. This may possibly be explained by the following reasons. In total in-house, food processors retain full control over the logistics operation and can respond quickly to customer needs (Beaumont and Sohal, 2004). Likewise, in total outsourcing, the LSPs (4PLs) take full control of the manufacturer's logistics operation from daily transportation to strategic planning activities such as relocation of warehouses or selection of road and ocean carriers. Through network optimisation, 4PLs could help manufacturers to cope with high demand complexity because 4PLs often have superior capabilities in assembling and managing multiple resources (Carbone and Stone, 2005). In addition, the literature also shows that the full logistics service provider often possesses a higher level of service capabilities than other types of LSPs, such as carrier or warehousing operators. For instance, they are better in making efforts to help in emergencies, responding flexibly to customer requests and recommending alternative actions when unforeseen problems arise (Lai, 2004). This may explain why total outsourcing is related to service performance.

Chapter 5. Comparisons of logistics outsourcing in the Taiwanese and Dutch food processing industries[16]

5.1 Introduction

The Chapter 2 through 4 investigated outsourcing determinants and the impact of logistics outsourcing on logistics performance. In the last part of this book, we look for the implications for the logistics industry. The main objective of this chapter is to compare the food processing industry's use of various logistics services in Taiwan and the Netherlands. This chapter aims to answer the fourth research question.

> *Research question 4*
>
> *RQ 4 What are the current and expected future developments in logistics outsourcing in Taiwan and the Netherlands?*

In order to answer this research question, three subquestions are formulated:

> *RQ 4a To what extent do levels of logistics outsourcing differ between Taiwan and the Netherlands in terms of current status and future plans?*
>
> *RQ 4b To what extent do outsourcing firms' characteristics differ between Taiwan and the Netherlands?*
>
> *RQ 4c What implications would any differences between Taiwan and the Netherlands have, in both logistics outsourcing levels and firms' characteristics, for the operational strategies of LSPs in Taiwan?*

The chapter is organised as follows. In Section 5.2, we present a review of the literature related to developments in the logistics industry and definitions of outsourcing of logistics activities. Subsequently, in Section 5.3 we outline the research method, and then follow this with the analysis of results in Section 5.4. Finally, Section 5.5 presents our discussions and conclusions.

5.2 Literature review

The process of logistics outsourcing is one that often involves the use of external logistics companies (third-party) to perform activities that have traditionally been performed within an organisation (Bagchi and Virum, 1996; Berglund *et al.*, 1999; Lieb, 2002; Londe and Cooper,

---

[16] This chapter is based on an article submitted to an international scientific journal: Hsiao, H.I., R.G.M. Kemp, J.G.A.J. van der Vorst, S.W.F. Omta, Logistics Outsourcing by Taiwanese and Dutch food processing industries.

1998; Sink and Langley, 1997). A third party is neither the seller (first party) nor the buyer (second party) in the supply chain. The term 'logistics company', 'logistics service provider' or 'outsourcer' is used to denote the firm that operates the logistics activities; and the term 'service user' or 'outsourcee' is used to denote the firm to whom the contract for services is given (Virum, 1993). In this section, we discuss logistics environments in the Netherlands and Taiwan, the development of the logistics industry, provide definitions of logistics outsourcing activities and review some prior studies on level of logistics outsourcing in Western and Asian countries.

5.2.1 Logistics environments in the Netherlands and Taiwan

Taiwan and the Netherlands are located centrally in their geographical regions, the Asia-Pacific rim and Europe. In terms of economic development and logistics environment, the Netherlands is well ahead of Taiwan (see Table 5.1). The Netherlands is known internationally as the logistics and distribution hub of Europe. A 2006 survey commissioned by Capgemini ranked the Netherlands as the most desirable location for European Distribution Centres (EDCs), especially in the high-tech and food and beverage sectors (Capgemini, 2006). According to the report, the Netherlands accounts for 51 percent of all European distribution centres within the EU market, with more than 9,000 foreign companies using the region as their distribution hub. Rotterdam is Europe's largest seaport and the main port for agribusiness. Each year, the port handles some 406 million metric tons of cargo (Port of Rotterdam, 2007). Schiphol is the third largest freight airport in Europe, with a total of 1.65 million tonnes (2007) of freight trans-shipment, after Frankfurt and Charles de Gaulle Airport in Paris. Taiwan trails

Table 5.1. Key social and logistics infrastructure indicators in Taiwan and the Netherlands (www.gio. gov.tw and www.cbs.nl).

	Taiwan	The Netherlands
Population (million)	22.82 (July 2006)	16.337 (May 2006)
Area (sq km)	36,200	41,526
GDP per capita	$29,500 (2006)	$35,078 (2006)
Economic growth rate (percent)	4.03 (2005)	2.9% (2006)
Unemployment (percent)	4.2 (2005)	5.5 (2006)
Highway length (km)	952	5,012
Rail network (km)	1,094	3,000
Airport (total)	18	21
Airports (international)	2	1
Key airport throughput (million tonnes)	1.60 (2007)	1.65 (2007)
Seaports (international)	7	13
Key seaport throughput (million tonnes)	146 (2007)	406 (2007)

the Netherlands in logistics development. Port of Kaohsiung is the largest seaport in Taiwan; however the port only handles some 146 million metric tons of cargo. Taoyuan Airport is the six largest freight airport in Asia-Pacific region, with only a total of 1.6 million tonnes of freight transhipment. Taiwan is a market-oriented economy and the supportive government policies have made it a highly competitive manufacturing and export base. Given the key role of efficient logistics services, the Taiwanese government has focused considerable attention on the development of the logistics environment. Developing Taiwan as an international logistics and distribution hub has become an important issue in the last few years(CEPD, 2002).

5.2.2 Development of the logistics industry

The literature on international logistics providers reveals three waves of entrants into the outsourcing market (Berglund *et al.*, 1999; Carbone and Stone, 2005). The first wave dates back to the 1980s or even earlier with the emergence of traditional logistics providers holding a strong position in either transportation or warehousing. In this wave, logistics services in the outsourcing market are the traditional services, such as transportation and warehousing.

The second wave dates from the early 1990s, with the arrival of a number of network players, for example DHL or TNT, who began providing their logistics services across a wider geographical area through various mergers and acquisitions. These activities increased substantially among both similar and contrasting types of players. For example, in 1996 Dutch TPG positioned itself among the world leaders of integrated logistics by acquiring TNT, the international integrator. Meanwhile, TNT itself had merged with a refrigerated transport specialist. Literature shows that mergers and acquisitions can achieve the following objectives (Berglund *et al.*, 1999; Carbone and Stone, 2005):
- wider geographical coverage and control of major traffic flows through efficient transport networks;
- economies of scope to improve operating margins through business process re-engineering and commercial entry into new market segments;
- strategic and operational synergies, through the acquisition of specialist capabilities, especially higher value-added services.

The third wave dates from the late 1990s when a number of players from the areas of information technology, management consultancy and financial services started working together with the LSPs from the first and second waves. The creation of partnerships among different players is led by the need to acquire competencies for the effective management of new and emerging customers (Carbone *et al.* 2005). The strategic or potential development options include supply chain management, combined intermodal transport and e-commerce. This period saw the introduction of a new service called the 'supply chain solution,' also known as fourth-party logistics (4PL) or leading logistics services because the new LSPs can lead traditional 3PLs to supply services to customers (Carbone and Stone, 2005; Hertz and Alfredsson, 2003).

In the next section, we divide the outsourceable logistics activities into four levels and discuss each in more detail.

5.2.3 Definition: level of logistics activities

Following developments in the logistics industry, the outsourceable logistics activities range from execution to planning (Dapiran *et al.*, 1996; Hong *et al.*, 2004; Lieb, 2002; Millen *et al.*, 1997; Razzaque and Sheng, 1998; Sahay and Mohan, 2006; Sohail *et al.*, 2006; Wilding and Juriado, 2004). In our study, the outsourceable logistics activities are divided into four levels (Hsiao *et al.*, 2006):

Execution activities:
- *Level 1*: includes transportation and warehousing. At this lowest level, contractual relationships between LSPs and their clients are often short term.
- *Level 2*: includes value-added activities, which refers to tasks normally performed by manufacturers but now being moved into distribution as part of final processing. In the food manufacturing industry, these tasks include mixing flavours, packaging or labelling. The contractual relationships between LSPs and their clients are often limited to one year or less.

Planning activities:
- *Level 3*: refers to the outsourcing of logistics planning and control activities, such as inventory management and transportation management. The LSPs offer customised logistics solutions and their skills are complementary to that of their clients.
- *Level 4* (total outsourcing): refers to outsourcing of the distribution network design or 4PL activities. When activities at this level are outsourced, the LSPs take care of all logistics activities, the logistics network design and orchestrate the logistics flow of the network (Van der Vorst *et al.*, 2007). At this strategic planning and control level, decisions are made concerning supply chain restructuring, for example, selection of road carriers, reassignment of roles and responsibilities among chain entities, changes of the warehouse structure, redistribution of inventory between tiers, changes in transportation network, mode, consolidation points.

5.2.4 Prior studies on logistics outsourcing in Western World and Asia

Here we review prior researches on logistics outsourcing in Western World and Asia. Table 5.2 presents usages information of logistics services in USA, Australia, Europe and Asia during the period 1996-2006. The countries of the Western World pioneered outsourcing logistics activities, in contrast to Asian countries, possibly because the logistics industry developed earlier in the West than in Asia. The Table 5.2 shows that in the early days, USA and Australia had already committed a large share of the relationship with logistics companies into a variety of logistics activities. For example, Lieb and Randall (1996) and Dapiran *et al.* (1996) report

Table 5.2 Usages of logistics service – Summary of academic works[a]

	USA, Australia and Europe			Asia		
	USA (Lieb, 2002)	Australia (Dapiran et al., 1996)	UK, France and Germany (Wilding and Juriado, 2004)	Singapore (Sohail et al., 2006)	Malaysia (2003)	India (Sahay and Mohan, 2006)
Sample	All industries	All industries	Consumer goods industry	All industries	All industries	All industries
Level 1	Shipment consolidation (33) Warehouse management (36)	Shipment consolidation (42) Warehouse management (47)	Transport (68) Storage (36)	Shipment consolidation (55.3)	Shipment consolidation 49 (58.3%)	Transportation (55.7) Warehousing (29.5)
Level 2	Product assembly (11)	Product assembly (13)	Re-labelling and re-packaging (40) Final product customisation (37)	Product assembly (6.6)	Product assembly/ installation 9 (10.7%)	Labelling and packaging (29.0) Assembly (12.7)
Level 3	Inventory replenishment (6)	Inventory replenishment (13)	-	Inventory replenishment (10.5)	Inventory replenishment 20 (23.8%)	Inventory management (23.8)
Level 4	-	-	-	-		-

[a] Percentages of surveyed firms that currently outsource this activity are provided in parentheses.

that in 1996 more than 33 of their respondents had outsourced level 1 activities in USA and Europe while more than 11 percent and 6 percent had outsourced product assembly (level 2) and inventory replenishment activities (level 3). If we compare the usage statistics of logistics services in a later period (2004-2006), we find greater usage in Europe than in Asia (e.g. India, Malaysia or Singapore). At level 1, for example, 68 percent of European respondents cited usage of transport (Wilding and Juriado, 2004), while only 55.7 percent of respondents in India cited usage of shipment consolidation (Sahay and Mohan, 2006). At level 2, 40 percent of respondents in Europe cited usage in re-labelling and re-packaging, in contrast to 29 percent of the respondents in India. It is pity that level 3 and 4 activities were not included in the cited studies, and were thus not available for comparison. However, based on this research, we predict that the Netherlands currently has a higher percentage of logistics outsourcing than Taiwan.

Asia is undergoing rapid economic expansion accompanied by growing regional trade and investment. Some studies have reported positive predicted growth of logistics outsourcing in Asia. For example, Aktas and Ulengin (2005) studied outsourcing logistics services in Turkey and conclude that it has great potential for further development. Sohail *et al.* (2006) also mention the wide scope for logistics services in future in Singapore and Malaysia based on current usage of contracted logistics services. Lieb (2008) surveyed the CEOs of ten logistics companies and asked them to identify the most significant opportunities available to logistics service providers in the Asia-Pacific outsourcing marketplace. Seven of those surveyed highlighted opportunities related to continued growth of the intra-Asian business and growth in the domestic markets of China and India.

5.3 Research method

Questionnaires were sent to Taiwanese and Dutch food processors with at least 40 employees. Lists of food companies were obtained from the Dutch Chamber of Commerce (www.kvk. nl) and Taiwan's Industry & Technology Intelligence Service (www.itis.org.tw). The five-page questionnaire was designed after consultation with colleagues, industry experts, and target respondents. The participating firms were first telephoned to obtain the name of the logistics manager. Within a week of the telephone contact, a questionnaire with a cover letter and pre-paid reply envelope was posted to the managers. Two weeks later, a first reminder, including another copy of the questionnaire and the cover letter and pre-paid reply envelope was sent to managers. The total sample population for this study was 890 (NL: 385; TW: 505); 66 questionnaires had incorrect address details and were returned by the postal service (NL: 57; TW: 9). A total of 138 responses were received (NL: 76; TW: 62), of which 24 had missing data (NL: 7; TW 17) and were judged unusable, thus yielding a sample size of 114 (NL: 69; TW: 45) with a response rate of 15% (NL: 21%; TW: 9%). This compares favourably with response rates of other studies on the use of logistics services (Dapiran *et al.*, 1996; Lieb, 2002).

The survey instrument focused on the following areas:
- level of logistics outsourcing (current and future);
- outsourcing firms' characteristics (firm size, logistics strategy and supply chain complexity).

Below we explain these measures in more detail.

Level of logistics outsourcing. Respondents were asked to describe the current outsourcing status of four levels of logistics activities: transportation (level 1), packaging; (level 2); transportation management (level 3); and distribution network design (level 4). The survey asked respondents to 'describe your current practice of logistics outsourcing in your main product group given the three options below.' These included 'we have already outsourced this activity,' 'we plan to outsource this activity,' and 'we don't intend to outsource this activity.'

Outsourcing firms' characteristics. Earlier studies showed that three types of firm characteristics are important for analysing differences in outsourcing decisions: firm size, logistics strategies and supply chain complexity. Take firm size as an illustration. Even with available funding for internal logistics activities, larger firms could also favor external alliances because they may have greater bargaining power (Robertson and Gatignon, 1998).
- Firm size is measured by number of employees.
- Literature distinguishes the following logistic strategies for food manufactures: cost reduction, reliability, flexibility, lead-time reduction, and food quality(/safety) (Beamon, 1999; Sum and Teo, 1999; Wheelwright, 1984). The instrument consisted of these five objectives, and respondents were asked to rank the importance of each objective as a percentage with the overall sum of 100.
- Supply chain complexity, from a logistical view, refers to the level and type of interactions present in the upstream and downstream logistics flow (Milgate, 2001). It is determined by its extent, variety (differentiated, inconsistent) and variability (fluctuation and predictability of changes) (Germain *et al.*, 2008; Hsiao *et al.*, 2008). A number of authors mention that the logistics outsourcing decision is positively related to supply chain complexity (Hsiao *et al.*, 2006; Milgate, 2001; Rao and Young, 1994). For example, a large number of suppliers increases the level of coordination needed to improve the efficiency of operations. With fewer suppliers, the focal company can implement a more efficient buyer-supplier interface through more cost-effective inventory control. In this regard, outsourcing of a certain logistics activity makes sense when a firm operates under high supply chain complexity, because the operational load to manage such a complex supply chain system could then decrease. The literature and logisticians suggest 17 supply chain complexity items which might complicate a food manufacture's logistics process (Ball, 2007; Hsiao *et al.*, 2008; Londe and Cooper, 1998; Stadtler, 2002; Van der Vorst and Beulens, 2002; Wanke and Zinn, 2004). Respondents were asked to rate the degree to which the item complicates logistics management in their product group on a seven-point Likert type scale ranging from (1) extremely low to (7) extremely high.

5.4 Results

This section presents comparative analyses of the findings on current outsourcing status and future plans. This section answers our initial research question: What are the most commonly outsourced logistics activities in Taiwan and the Netherlands?

5.4.1 Level of logistics outsourcing in Taiwan and the Netherlands

Table 5.3 exhibits the results for the current extent of outsourcing in Taiwan and the Netherlands. Level 1 activities are the most commonly outsourced logistics activities in the Netherlands and Taiwan. About 69% of the companies in both countries outsource level 1 activities, 16% level 2 and 37% level 3 activities. Only few companies (about 10%) outsource the highest level of activities. In particular, the Netherlands has higher percentages for levels 1 and 3. Table 5.4 shows the results of 'planning to outsource' in the future for all levels of activities, and Table 5.5 shows the results of 'won't outsource' in the future. When intentions for the future are included, Taiwan will outsource level 2 (40%) and 4 activities (36%) much more than the Netherlands (resp. 13% and 17%). These results are roughly in line with predictions for other Asian countries such as Turkey (Aktas and Ulengin, 2005), Singapore and Malaysia (Sohail *et al.* 2006).

5.4.2 Differences between Taiwanese and Dutch outsourcing firms

This section answers our second research question: *To what extent do outsourcing firms' characteristics differ between Taiwan and the Netherlands?* Firm size, logistics strategy and supply chain complexity were studied to analyse differences in Dutch and Taiwanese outsourcing firms. Table 5.6 presents a detailed distribution of the current outsourcing firms in the four levels of activities and in eleven food (sub)sectors in Taiwan and the Netherlands. Sector categorisation was based on the standard categorisation of CBS. The table shows that

Table 5.3. Frequencies cited by respondents in each level of activity: comparing 'had outsourced' in Taiwan (TW) and the Netherlands (NL)[a].

	'had outsourced'		
	Total (N=114)	TW (N=45)	NL (N=69)
Level 1: transportation	79 (69%)	27 (60%)	52 (75%)
Level 2: packaging	18 (16%)	9 (20%)	9 (13%)
Level 3: transportation management	42 (37%)	14 (31%)	28 (41%)
Level 4: distribution network design	12 (11%)	5 (11%)	7 (10%)

[a] Percentage provided in parentheses refers to frequency relative to sample population.

Table 5.4. Frequencies cited by respondents in each level of activity: comparing 'plan to outsource' in Taiwan (TW) and the Netherlands (NL)[a].

	'plan to outsource'		
	Total (N=114)	**TW (N=45)**	**NL (N=69)**
Level 1: transportation	10 (9%)	6 (13%)	4 (6)
Level 2: packaging	9 (8%)	9 (20%)	-
Level 3: transportation management	9 (8%)	5 (11%)	4 (6%)
Level 4: distribution network design	16 (14%)	11 (25%)	5 (7%)

[a] Percentage provided in parentheses refers to frequency relative to sample population.

Table 5.5. Frequencies cited by respondents in each level of activity: comparing 'won't outsource' in Taiwan (TW) and the Netherlands (NL)[a].

	'won't outsource'		
	Total (N=114)	**TW (N=45)**	**NL (N=69)**
Level 1: transportation	25 (22%)	12 (27%)	13 (19%)
Level 2: packaging	87 (76%)	27 (60%)	60 (87%)
Level 3: transportation management	63 (55%)	26 (58%)	37 (53%)
Level 4: distribution network design	86 (75%)	29 (64%)	57 (83%)

[a] Percentage provided in parentheses refers to frequency relative to sample population.

there are significant differences between the countries and the subsectors; further analyses also showed that the firm's characteristics are different between the sectors. Therefore, we will focus in the remainder of this article on two large and representative sectors: the meat and dairy sectors. Furthermore, we focus on level 1 and 3 chosen as examples for analysis of differences since they represent relevant sample sizes.

When we compare the outsourcing levels in the meat and dairy sectors in both countries, we find the following differences. For the meat sector, the Netherlands have higher scores in level 1 and 3; Taiwan scores higher for level 2 and 4. The dairy sector as whole scores higher on outsourcing on the first three levels, none of the companies outsources level 4. Also here the Netherlands have more outsourcing in level 1 and 3, and Taiwan more in level 2. Compared to the meat sector, the dairy sector as whole scores higher on outsourcing on the first three

Table 5.6. Frequencies cited by outsourcing respondents in Taiwan (TW) and the Netherlands (NL) in each level of activity and in each category of food sector[a].

Sector	TW					NL				
	Total (N=45)	Outsourcing firms				Total (N=69)	Outsourcing firms			
		L1	L2	L3	L4		L1	L2	L3	L4
Meat	7	4 (57%)	1 (14%)	1 (14%)	2 (29%)	12	9 (75%)	-	3 (25%)	-
Fish	3	2 (67%)	1 (33%)	1 (33%)	-	2	2 (100%)	-	-	-
Fruit/veg	4	2 (50%)	-	2 (50%)	-	6	5 (83%)	1 (17%)	3 (50%)	1 (17%)
Oils/fats	-	-	-	-	-	2	1 (50%)	-	-	-
Dairy	4	3 (75%)	1 (25%)	1 (25%)	-	8	8 (100%)	1 (13%)	5 (63%)	-
Grain mill	5	5 (100%)	1 (20%)	2 (40%)	-	1	1 (100%)	1 (100%)	1 (100%)	1 (100%)
Animal feed	3	1 (33%)	1 (33%)	-	-	8	3 (38%)	-	-	-
Others	6	3 (50%)	1 (17%)	2 (33%)	2 (33%)	27	20 (74%)	4 (15%)	13 (48%)	4 (15%)
Beverages	6	5 (83%)	3 (50%)	2 (33%)	-	2	2 (100%)	1 (50%)	2 (100%)	1 (50%)
Lunch box	5	1 (20%)	-	1 (20%)	-	-	-	-	-	-
Prepared meal	2	1 (50%)	-	1 (50%)	1 (50%)	-	-	-	-	-

[a] Percentage provided in parentheses refers to frequency relative to sample population (N) of sector. Note that a firm may outsource activities at multiple levels; numbers may therefore not add up to the total sample size.

levels. Let us now look at the firms' characteristics in more detail and see if we can explain these differences.

(1) Firm size

Firm sizes, i.e. the number of employees, of the outsourcing respondents from the two countries are presented in Table 5.7. We may conclude that there are differences between both countries, but no differences between the sectors. The dairy sector in the Netherlands has more smaller firms than the dairy sector in Taiwan (for example, percentage of dairy-TW-L1 in 250+ is larger than percentage of dairy-NL-L1). Also the meat sector in Taiwan represents firms with many employees, whereas the Netherlands represents smaller companies (for example, percentage of meat-TW-L1 in 150-<250 is larger than percentage of meat-NL-L1). This is probably also due to the fact that manual labour is much cheaper in Taiwan than in the Netherlands. This could explain why companies in Taiwan outsource value-adding activities more than the Netherlands; there is an available low-cost work force.

(2) Logistics strategy

Respondents were also asked to rank the importance of five logistics objectives as percentages with an overall sum of 100. Figure 5.1 presents the results.

From the figure we can conclude that the outsourcing strategies of companies in the subsectors and in the countries differ. It turns out that in both countries food quality, reliability and costs are the most important performance indicators at the lowest level for both sectors; however, Taiwan emphasises low cost whereas the Netherlands includes flexibility as a major issue. At higher outsourcing levels, we see differences in country and sector; low costs become less

Table 5.7. Firm size (number of employees) of Taiwanese (TW) and Dutch (NL) outsourcing firms in the dairy and meat categories and in the L1 and L3 outsourcing categories.

Employees	Dairy				Meat			
	TW		NL		TW		NL	
	L1	L3	L1	L3	L1	L3	L1	L3
<50	0%	0%	37%	20%	0%	0%	22%	0%
50-<100	0%	0%	13%	20%	25%	0%	45%	100%
100-<150	0%	0%	25%	20%	25%	50%	22%	0%
150-<250	33%	0%	25%	40%	50%	50%	11%	0%
250+	67%	100%	0%	0%	0%	0%	0%	0%

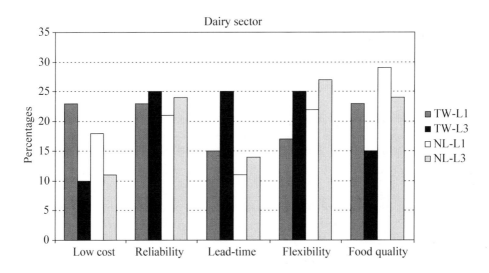

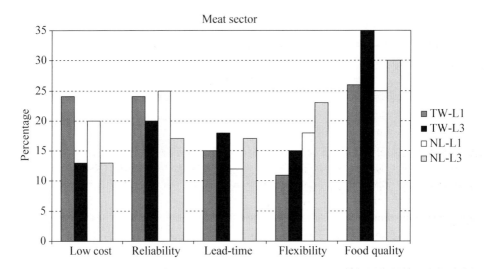

Figure 5.1. Logistics strategy of Taiwanese (TW) and Dutch (NL) outsourcing firms in the dairy and meat categories and in the L1 and L3 outsourcing categories.

important and flexibility becomes more important. In general, meat companies emphasise food quality, and next to that the Netherlands aims for flexibility whereas Taiwan focuses on reliability and food quality. In general, dairy companies emphasise flexibility and reliability, next to that the Netherlands aims for food quality and Taiwan for lead-time improvement. We may conclude that the Netherlands aims for higher flexibility objectives in both levels. Taiwan focuses more on the basic performance indicators as low costs and lead time improvement.

This might be explained by the complexity factors. Further discussions will be given in the concluding remarks.

(3) Supply chain complexity

Respondents were asked to rank to what degree a certain supply chain characteristic complicated their logistics management. Figure 5.2 (for level 1) and 5.3 (for level 3) present the results for the averages of the seventeen complexities items. The figures show major differences in supply chain complexity factors that may also explain differences in outsourcing decisions.

Findings level 1

In the meat sector, the average supply chain complexities of the Taiwanese meat outsourcing firms is 3.8 (see L1-meat-TW) and the Dutch meat outsourcing firms is 4.2 (L1-meat-NL). Dutch meat outsourcing firms have some higher complexities in the number of international customers. Taiwanese meat firms have some higher complexities in storage variety.

In the dairy sector, the average supply chain complexity of the Taiwanese dairy outsourcing firms is 3.5 (L1-dairy-TW); and the Dutch dairy outsourcing firms is 4.4 (L1-dairy-NL). Dutch dairy outsourcing firms have some higher complexities in: number of products, demand fluctuation, number of customers, number of international customers, and delivery frequency.

Findings level 3

Interestingly in the meat sector, the average supply chain complexity of Dutch meat outsourcing firms was lower (3.8) than in Taiwan (4.9) (see L3-meat-TW and L3-meat-NL. Dutch meat outsourcing firms have some higher complexities in the following items: number of international customers; the Taiwanese meat outsourcing firms have significant higher complexities in most of the other factors, especially for the number of product groups, storage variety, distribution channel variety.

In the dairy sector, the average supply chain complexity of the Dutch dairy outsourcing firms was 5.0 compared to 3.7 for the Taiwanese firm (see L3-dairy-TW and L3-dairy-NL). The Dutch dairy outsourcing firms have some higher complexities in the following items: production uncertainty, demand fluctuation, number of international customers, distribution channel variety, and delivery frequency.

In summary, from the figures we may conclude that supply chain complexity of companies in the subsectors and in the countries differs. Both at high and low levels, Dutch firms have higher complexities than Taiwanese firms. Besides, Taiwanese meat firms have higher complexities than dairy firms, while Dutch dairy have higher complexities than meat firms. The figures also

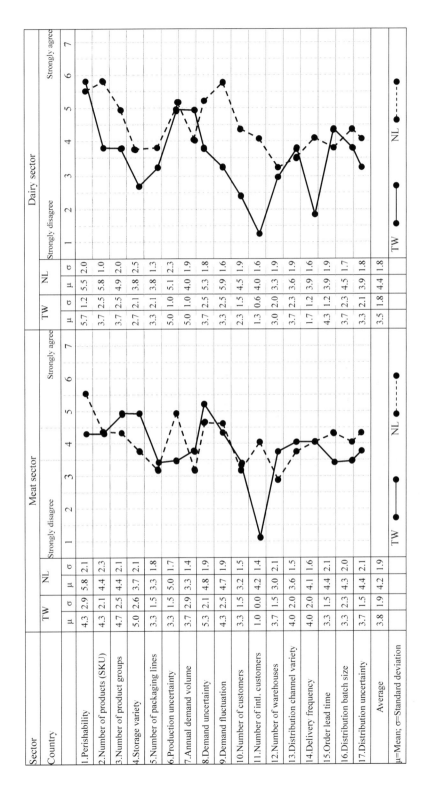

Figure 5.2. Level 1 outsourcing: supply chain complexity of Taiwanese (TW) and Dutch (NL) outsourcing firms in the meat and dairy categories.

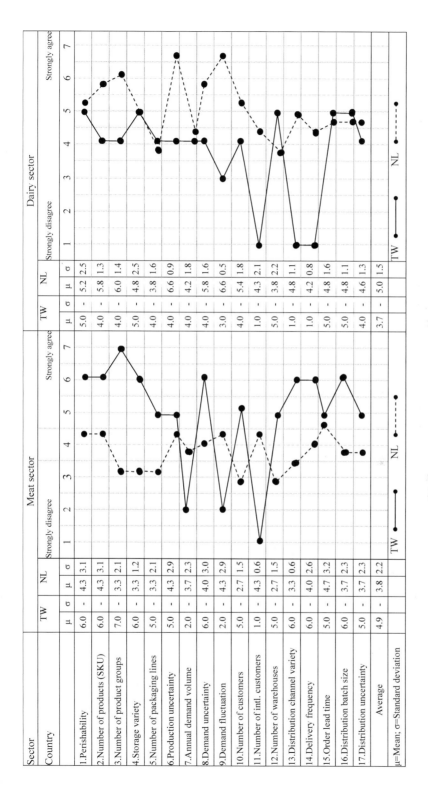

Figure 5.3. Level 3 outsourcing: supply chain complexity of Taiwanese (TW) and Dutch (NL) outsourcing firms in the meat and dairy categories.

indicate that product perishability, delivery frequencies, distribution channel variety, but also demand fluctuation and uncertainty differ between the sectors.

Concluding results

From the previous survey results we may conclude the following:
- Most companies outsource level 1 activities, especially in the Netherlands, with low cost, food quality and reliability as the most important performance indicators; however, Taiwanese firms emphasise low cost whereas Dutch firms also focus on flexibility. This is mainly due to the fact that – at this level - the Netherlands has to deal with higher complexities regarding the number of international customers and distribution requirements. Outsourcing transportation opens up markets by providing wider geographical coverage and control of major traffic flows through efficient transport networks, which cannot be operated by the smaller Dutch food companies.
- Dutch firms have relatively higher outsourcing percentages for level 1 and level 3 activities than Taiwanese firms; Taiwanese firms have relatively higher percentages for level 2 activities than Dutch firms. In the future, the level of outsourcing on the level 2, 3 and 4 is expected to increase in Taiwan, overtaking the Netherlands. This might be caused by: (1) the abundant availability of cheap manual labour in Taiwan, which provides room for low-cost activities but requires outsourcing for a range of control reasons; (2) in future, the complexity may increase, for example, in the meat industry, resulting in a need to reduce that complexity by outsourcing certain activities; (3) high pressures on cost reduction (and still less emphasis on service, although increasing), which requires full truck utilisation and optimal distribution network design; (4) increasing service requirements from customers combined with increasing customer demands on lead time reduction, reliability and flexibility improvements.
- Sectors differ in firm characteristics and logistics strategies which impacts their decision making on outsourcing logistics activities. We found that, compared to the meat sector, the dairy sector as a whole scores higher on outsourcing on the first three levels. This might be caused by the following: (1) meat companies place greater emphasis on food quality, but dairy companies focus more on flexibility and reliability; (2) Figure 5.2 and 5.3 indicate that product perishability, delivery frequencies, distribution channel variety, but also demand fluctuation and uncertainty differ between the sectors. This combined with differences in customer requirements results in different outsourcing strategies.

5.5 Discussion and conclusions

In this section, we begin by drawing conclusions related to the first two research questions, which considered logistics outsourcing levels and strategy in Taiwan and the Netherlands. Next, based on these conclusions, we derive some implications for LSPs, thereby addressing the final research question.

This study aimed to compare food industry outsourcing of various activities in Taiwan and the Netherlands. It delivers some interesting results. First, our findings show that the most commonly outsourced activities in the Netherlands and Taiwan were the level 1 activities. About 69% of the companies in both countries outsource level 1 activities, 16% level 2 and 37% level 3 activities. Only a few companies (about 10%) outsource level 4 activities. In particular, the Netherlands has higher percentages for levels 1 and 3. When outsourcing intentions are included, Taiwan will outsource level 2 (40%) and 4 activities (36%) much more than the Netherlands (resp. 13% and 17%). When zooming in on the subsectors, it turns out that outsourcing strategies of companies in these subsectors differ. For instance, the dairy sector outsources more frequently than the meat sector on the first three levels.

Second, regarding outsourcing firms' characteristics, we found that logistics strategy and supply chain complexity were related to the outsourcing levels. Dutch firms aim for higher flexibility. Taiwanese firms focus more on low costs. Furthermore, most of the Dutch firms operated under higher supply chain complexities than Taiwanese firms. We contend that these differences may explain why Dutch firms had relatively higher percentages for level 1 and 3 outsourcing than Taiwanese firms. Most Taiwanese companies emphasise low cost whereas the Netherlands also focuses on flexibility. This is mainly due to the fact that the Netherlands has to deal with higher complexities regarding number of international customers and distribution requirements.(Hsiao *et al.*, 2008; Milgate, 2001; Rao and Young, 1994). In addition, we found that meat companies place greater emphasis on food quality, but dairy companies focus more on flexibility and reliability. These differences may explain why meat firms had relatively lower percentages for the first three levels than dairy.

Operations strategies and implications for LSPs in Taiwan

Here, we answer our third research question: *What implications would any differences between Taiwan and the Netherlands have, in both logistics outsourcing levels and outsourcing firms' characteristics, for the operational strategies of LSPs in Taiwan?* First, we consider the implications for local LSPs, then the implications for international LSPs.

(1) Local LSPs. Our findings reveal that the outsourcing of level 2, 3 and 4 activities can be expected to increase in the future in Taiwan. Furthermore, we also found that the high level of outsourcing firms tended to focus more on flexibility, the firms that only outsource level 1 logistic activities tended to focus more on low cost. Although outsourcing users will increase, at the moment it is hard to find local LSPs that can provide a wide range of services comparable to those in the Dutch logistics industry. Many local LSPs provide basic services, such as shipping or transportation, and are struggling to survive in tough competitive markets. This is likely because a pure cost posture is easier to replicate by potential new entrants (Wang 2006). Therefore, we suggest that pure cost companies review their current strategy and decide if they want to continue the pure cost strategy with low cost and low profit or migrate to becoming a differentiation-oriented provider in order to provide more services to their customers in the

form of more flexibility and reliability, amongst other things. After the entry of Chinese Taipei (Taiwan) to the World Trade Organisation in 2002, international LSPs can now enter the Taiwanese service markets with tremendous advantages of capital, technology and operations experience. It will be difficult for local LSPs to compete with these giants if they cannot offer similarly innovative services of high service quality. Thus, we suggest that local cost-driven LSPs will in coming years transform themselves into service-oriented providers offering a wide variety of services.

(2) International LSPs. Our findings show that Taiwanese and Dutch firms employ different strategies. This result suggests that international LSPs should not use a uniform strategy when entering different countries (Cullen and Parboteeah, 2005). When entering a new region, the LSP should realise that potential customers may have fundamentally different needs than the provider's existing customers in the home region (Sink, 1996; Arryo 2006).

Finally, the outsourcing market for value-added and high-level logistics activities shows good potential for further development in Taiwan. However, Taiwan's logistics service providers still have some problems:

- A shortage of food safety and quality personnel. This is particularly important for LSPs with large clients in the food industry. The implication for the LSPs also comes from the differences between meat and dairy outsourcing firms. We found that the meat sector focus more on food quality than the dairy sector. Various types of food products require different preservation knowledge. Taiwan trails the Netherlands by at least a decade when it comes to the implementation of food safety and control systems. Hazard analysis and critical control points (HACCP) is a food safety management system recognised worldwide. Dutch transport and distribution sectors have been required to implement HACCP since 2006. Although HACCP was introduced to Taiwan in 1997, it is still a voluntary measure for Taiwanese food and service industries (Jeng and Fang, 2003). Therefore, implementing a HACCP system would be one good method of improving service quality in Taiwanese logistics firms.
- Infrastructure. Traffic congestion is an obstacle to the growth of logistics. The loading and unloading of cargos during normal working hours is considered the main cause of congestion in urban areas. Lack of parking space and traffic congestion in urban areas leads to increasing transport costs (Feng and Chia, 2000).

A shortage of qualified logistics personnel is another problem. Taiwan's logistics industry still lacks operation research (OR)-related organisational units. OR studies cover issues such as linear programming, transportation problems, network analysis, and queuing models. The fact that companies are unfamiliar with OR techniques and there is an additional shortage of OR professionals will become a problem for the further development of the logistics industry (Bremmers *et al.*, 2004).

Chapter 6. Discussion and conclusions

In this chapter, we will discuss the main findings of the different studies presented in this book. In Section 6.1 we briefly outline our research. In Section 6.2, we summarise the most important findings. Section 6.3 discusses main contributions to literature. Section 6.4 focuses the research limitation and suggests future researches. Section 6.5 draws managerial implications. Finally, this chapter closes with some final remarks.

6.1 Brief outline of the research

The overall objective of this book is *to analyse how food processors determine their logistics outsourcing need and to analyse how logistics outsourcing influences logistics performance*. The study was undertaken in the Netherlands and Taiwan. In order to realise the objective, four research questions were formulated.
1. What kind of logistics activities can be outsourced by food processors? (1a) and what decision criteria are considered when outsourcing logistics activities? (1b)
2. What decision-making criteria are considered by food processors when outsourcing a certain level of logistics activities?
3. What is the impact of logistics outsourcing on service performance?
4. What are the current and expected future developments in logistics outsourcing in Taiwan and the Netherlands?

Chapter 2 was primarily concerned with building the decision-making framework and aimed to answer research question 1. The research design comprised three stages. A literature review was undertaken to study outsourcing theories and identify outsourceable logistics activities. Successively, case studies on three Dutch food processors were conducted resulting in a framework for make-or-buy decisions. Finally, an exploratory survey was undertaken in the Netherlands to examine the factors that determine the outsourcing decisions of different logistic activities.

Chapter 3 tested the decision-making framework using data from a mailed survey collected in the Netherlands and Taiwan and aimed to answer research question 2. Surveys were mailed to logistics managers in the food processing industry of companies with at least forty employees. Of the 890 questionnaires mailed (NL: 385; TW: 505), 66 had incorrect contact information (NL: 57; TW: 9) and were returned by the postal service. A total of 138 responses were received (NL: 76; TW: 62), of which 24 had missing data (NL: 7; TW: 17) and were judged unusable, thus yielding a sample size of 114 (NL: 69; TW: 45), a usable response rate of 15% (114/800) (NL: 21%; TW: 9%).

Chapter 4 was concerned with the impact of logistics outsourcing on logistics performance and aimed to answer research question 3. The outsourcing of four levels of logistics activities were examined: the transportation (Level 1), packaging (Level 2), transportation management

(Level 3), and distribution network design (Level 4). A research framework was formulated to discuss the effect of the outsourcing decision on perceived logistics service performance and includes the moderating role that supply chain complexity may play in the proposed relationships. The performance framework was tested by using a mailed survey collected from food processing industry in the Netherlands and Taiwan.

Chapter 5 focused on the implications for the LSPs. This chapter investigated the current and expected future developments in logistics outsourcing in the Netherlands and Taiwan and aimed to answer research question 4. The survey was used to evaluate the most commonly outsourced activities and identify specific outsourcing firms' characteristics.

6.2 Main findings and conclusions

Below we summarise our main findings for the four research questions.

RQ 1a What kind of logistics activities can be outsourced by food processors?

An extended literature analysis resulted in four levels of outsourceable logistics activities: *Level 1* refers to the execution level of basic activities, such as transportation and warehousing. *Level 2* refers to value-added activities. In food industry, packaging or labelling are examples. *Level 3* refers to the planning and control level. Activities that can be outsourced at this level are inventory management and transportation management. Sub-activities of inventory management are sales forecasting, stock control and event control. Sub-activities of transportation management include route planning and scheduling and event control. *Level 4* refers to the distribution network design. This is the strategic planning and control level in which decisions are made concerning road carrier selection, location and site analysis and logistics network management. When activities at this level are outsourced, the LSP takes care of the logistics network design and orchestrates all logistics flows in the network.

RQ 1b What decision criteria are considered when outsourcing logistics activities?

The literature analyses resulted in a decision-making framework (see Figure 6.1). The framework is based upon three theories: transaction cost theory, resource-based theory and supply chain management. Our exploratory case studies and survey suggested five important factors:

- Asset specificity refers to logistics-asset specificity. Logistics-specific assets involve investments in physical capital which will lose value if they are redeployed for other uses.
- Performance measurement uncertainty refers to the degree of difficulty associated with assessing the performance of transaction partners (the logistics service providers).
- Core (business) closeness refers to logistics capabilities, skills and/or experiences, with which a firm could gain greater value than competitors.

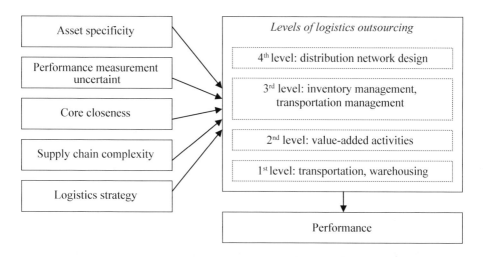

Figure 6.1. A decision making framework for level of logistics outsourcing in the food processing industry.

- Supply chain complexity refers to the number of elements within the focal company's logistical flow (on bases of production, distribution and demand), and the degree to which these bases are differentiated or varied. It influences the effort or operational load to manage the firm's logistical system.
- Logistics strategy includes three dimensions: (1) cost, (2) flexibility, and (3) food quality. Low cost strategy refers to a strategy in which companies seek to design logistics system more cost-efficiently than its competitors. Flexibility strategy refers to a strategy in which companies aimed at being flexible to the changing and diverse needs of customers. Food quality strategy refers to a strategy in which companies aimed at providing freshness, low damage and high food quality of a food product.

> *RQ2 What decision-making criteria are considered by food processors when outsourcing a certain level of logistics activities?*

This study was conducted in Taiwan and the Netherlands. The relationships between outsourcing levels and asset specificity, performance measurement uncertainty, core closeness and supply chain complexity were tested using binary logistic regression, as discussed in Chapter 3; the relationships between levels and logistics strategy were examined qualitatively, as discussed in Chapter 5. Table 6.1 present the propositions and the results of our hypotheses testing. Based on the analysis, we conclude the following for each factor:

- *Asset specificity* determines the outsourcing decision in the first three levels; the proposition was not supported for the Level 4 activities. We speculated that there are fewer specific investments to be made in the strategic planning level, thus asset specificity is less related to outsourcing decisions of the highest level.

Table 6.1. summary of propositions.

Propositions			
	Level 1	Level 2	Level 3	Level 4
P1: The higher the asset specificity of a specific logistics activity, the less likely that a food processor will outsource this activity than keep it in-house.	Confirmed	Confirmed	Confirmed	Not confirmed
P2: The higher the performance measuring uncertainty when outsourcing a logistics activity, the less likely that a food processor will outsource this activity than keep it in-house.	Not confirmed	Not confirmed	Not confirmed	Not confirmed
P3: The closer a logistics activity is to the core business, the less likely that a food processor will outsource this activity than keep it in-house.	Rejected	Rejected	Confirmed	Not confirmed
P4: The higher the supply chain complexity, the more likely a food processor will outsource a logistics activity than keep it in-house.	Not confirmed	*Distribution complexity:* confirmed	Not confirmed	*Demand complexity:* confirmed
P5: A food processor with a low cost strategy is more likely to outsource a logistics activity than a food processor with a flexibility or food quality strategy.	Confirmed	(was not tested)	Rejected	(was not tested)

- *Performance measurement uncertainty* was not confirmed for all levels. Poppo and Zenger (1998) also found no support for performance measurement difficulty on outsourcing decisions. They argue that because measurement accuracy has equivalent effects internally and externally, outsourcing choices will hinge on other factors. When managers can not easily measure the performance of an outsourced activity, they are less satisfied with its cost. Likewise, when managers can not easily measure performance of internal activities, they are also less satisfied with the performance of internal activities. These arguments may explain why in our study the performance measurement uncertainty was not confirmed for all levels.

- *Core closeness* determines the level 3 outsourcing decision. The proposition was rejected for all execution level of activities (Level 1 and 2) and not confirmed for level 4. We speculate that LSPs have provided the basic services since the early 1980s and may have developed very good capabilities. Therefore, outsourcing these activities might be the preferred choice for food processing companies although they regard these activities as important and central to core business.

- *Supply chain complexity* was confirmed for Level 2 regarding distribution complexity (numerous packaging lines, numerous clients, high delivery frequency and short lead-time) and confirmed for Level 4 regarding demand complexity (high demand volume, high demand uncertainty and high demand fluctuation). This result was consistent with expectations and findings in previous literature. A number of authors contend that companies attempted to achieve higher level of integration especially when they faced high levels of uncertainties (Van Donk and Van der Vaart, 2005; Wong and Boon-itt, 2008).

- *Logistics strategy* is related to the outsourcing decision. Findings in Chapter 5 show that a food processor with a low-cost strategy is more likely to outsource Level 1 activities. A food processor with a flexibility strategy is more likely to outsource Level 3 activities. Food quality strategy seems to function as a moderator which decreases the outsourcing probability in both levels. It seems that a firm with a food quality strategy is less likely to outsource an activity no matter whether the firm operates under low or high supply chain complexity. This may be explained by the fact that food companies that emphasise food quality usually require high standards on food quality management. Such companies are unlikely to outsource logistics activities because of their concern that LSPs are often unable to meet their standards and requirements.

Based on the above findings, we conclude the following outsourcing criteria for the different levels:

- In Level 1 activities, asset specificity, core closeness and low cost strategy are the decisive factors for a food processor.

- In Level 2 activities, asset specificity, core closeness and the distribution complexity (characterised by number of packaging lines, number of clients, delivery frequency and lead-time) are the decisive factors.

- In Level 3 activities, asset specificity, core closeness and flexibility strategy are the decisive factors for a food processor.

- In Level 4 activities, demand complexity (characterised by demand volume, demand uncertainty and demand fluctuation) is the decisive factor.

RQ3 What is the impact of logistics outsourcing on service performance?

In Chapter 4 we sought to advance our understanding of relationships between the outsourcing decision, the outsourcing level and a firm's logistics service performance. Two propositions were formulated and tested (Table 6.2).

Our findings show that most of the relationships between outsourcing and service performance are not confirmed. This indicates that there is no direct outsourcing effect on service performance. This may be roughly in line with some other market surveys, which often indicate that the main problem with LSPs is a lack of service level improvement (Capgemini, 2007; Wilding and Juriado, 2004).

However, proposition P2 was confirmed in level 4 which shows the relationship between logistics outsourcing and service performance is moderated by demand complexity at level 4. This indicates that service performance of outsourcing at level 4 increases with an increasing degree of demand complexity. One possible interpretation is that in highly unpredictable and complicated environment, food processors tend to select LSPs with high service capabilities to cope with such uncertain situations. In addition, LSPs might tend to specialise in developing particular technologies or processes to provide innovative solutions for food processors. Thus, an unpredictable environment may increase the service benefits of level 4 outsourcing.

RQ 4 What are the current and expected future development in logistics outsourcing in the Netherlands and Taiwan?

About 69% of the companies in both countries outsource Level 1 activities, 16% Level 2 and 37% Level 3 activities. Only few companies (about 10%) outsource the highest level of activities. In particular, the Netherlands has higher percentages for levels 1 and 3. This might be caused by the fact that most Taiwanese companies emphasise low cost whereas the Dutch

Table 6.2. summary of propositions (performance).

	Propositions
PI (perf.): Logistics outsourcing has a positive effect on a firm's logistics service performance.	Not confirmed at all levels
P2 (perf.): Firms operating in a supply chain setting with high logistical complexity gain greater logistics service performance benefits from logistics outsourcing.	Confirmed at the level 4

companies focus on flexibility in order to deal with higher complexities. When intentions for the future are included, Taiwan is planning to outsource Level 2 (40%) and 4 activities (36%) much more than the Netherlands (resp. 13% and 17%). When zooming in, we found that outsourcing strategies of companies in the subsectors differ. For instance, the dairy sector outsources more frequently than the meat sector on the first three levels. This might be caused by the fact that meat companies emphasise food quality, whereas dairy companies emphasise flexibility and reliability.

6.3 Theoretical contributions

This research has made four main contributions to literature.

First, this book has integrated different theories into one comprehensive decision-making framework. This research has put extensive effort into developing a conceptual framework that integrates transaction cost theory, resource-based theory and supply chain management insights in one overall model, as discussed in Chapter 2 and 3. We have gathered strong indications, both theoretical and based on empirical evidence, that such an integrative approach is needed to grasp the complexity of the outsourcing decision. For example, transaction cost theory explains more about the lower level of outsourcing, while supply chain management explains more about the higher level of outsourcing. These different theoretical perspectives deal with partly overlapping phenomena in complementary ways. This is in line with recent development and strategic outsourcing literature calling for a broader and multidimensional approach to outsourcing decision issues (Holcomb and Hitt, 2007; Jacobides and Hitt, 2005; Madhok, 2002).

Second, we analysed the outsourcing decision for different levels of logistics activities. No studies investigate outsourcing decisions for these levels of logistics activities in an integrated way. Our findings indicate that each level of activity has its own outsourcing considerations. For example, asset specificity is important for first three levels but not important for Level 4. Supply chain complexity is important for Level 2 and 4, but not for Level 1 and 3. Thus, we suggest other outsourcing researches analysing activities at different activities and levels.

Third, we included supply chain complexity as a moderating factor to test the relationship between the level of logistics outsourcing and service performance. To our knowledge, no empirical research has been done to test this moderating effect on logistics outsourcing and service performance. We have shown that logistics outsourcing has no direct impact on service performance; but that service performance only increases under certain conditions, such as high demand volume, high demand uncertainty and high demand fluctuation. The relationship between logistics outsourcing and firms' service performance is more complex than it appears. Including supply chain complexity as a moderator is in line with the increasing focus on environmental dynamism and outsourcing performance (Gilley *et al.*, 2004; Gilley and Rasheed, 2000; Salimath *et al.*, 2008).

Fourth, we applied insights with respect to logistics outsourcing in the food processing industry. The food processing industry has some special characteristics. Food deteriorates easily and food storage and transport requires investments in cooling and hygiene equipments. In addition, not much research on outsourcing has been carried out in the food industry. Despite the barriers to our research we were able to propose a theory-based model which explains outsourcing considerations for food processors. Therefore, the present study takes an important step in enlarging the body of knowledge on the logistics outsourcing in the food industry (Bourlakis and Weightman, 2004).

6.4 Research limitations and further researches

This research provides a detailed look at the determination of logistics outsourcing decisions by food processors and investigates its impact on performance, but it was limited in several ways that might be addressed in future research.

First of all, the number of firms that had outsourced Level 4 activities was relatively low. This might be due to the fact that this outsourcing service is still new to the market. We encourage further research to investigate other developed countries, such as USA or Canada and other industries that are expected to have more outsourcing cases in Level 4.

Second, we examined the direct impact of levels of outsourcing and moderating impact of supply chain complexity on service performance. This is a causal phenomenon. To test the propositions/hypotheses, however, we used cross-sectional data, as discussed in Chapter 3. Ideally, we would like to use longitudinal data, i.e. measuring the independent variable service performance at a later moment in time than the independent variables. Further research is encouraged to investigate the longitudinal relations in a quantitative approach. Given the cross-sectional setup of our data, our findings concerning causal relationships should be interpreted with some caution.

Third, this study generalises and compares results for different food subsectors, as discussed in Chapter 5. We suggest that outsourcing decisions might vary between food subsectors. In our study, we only selected meat and dairy processors due to the available number of representative cases in each level of logistics activity in these subsectors. For statistical reasons, it was not possible for us to divide our sample into eleven food subsectors. It may be interesting to see if the outsourcing behaviours are different between different food subsectors. Therefore, further research is needed to replicate this study in order to have a larger sample size in each subsector, or to use the case study method.

Fourth, we proposed a logistics outsourcing framework for food processors, as shown in Chapter 2. However we excluded logistics strategy from binary logistic regression test in Chapter 3; we discussed the effect of logistics strategy on logistics outsourcing qualitatively in Chapter 5. This is due to the fact that we have limited number of observations collected from this study. With a small sample, the accuracy of prediction is much less in large sample (Hair 2008). Therefore, further research is suggested to replicate this study in different settings with a large sample size.

6.5 Managerial implications

This section contains a translation of our empirical results into practical implications for decision makers. We formulate several important implications for the food processors and logistics service providers. Below we provide managerial suggestions based on our research to answer the following questions: *If a food processor wants to make a logistics outsourcing decision, what factors should be taken into account? If a logistics service provider wants to include food processors as major customers, what special factors should be taken into account and what are the opportunities?*

To food processors

First, carefully select the logistics activities to be outsourced

Outsourcing could help companies to achieve outstanding performance; however, some researchers also propose that the improper use of outsourcing could play an important role in the competitive decline of firms (Jenning (Barthelemy, 2003; Brandes *et al.*, 1997; Jennings, 1997). Our results support their viewpoints and suggest that outsourcing different logistics activities requires different decisive considerations. Before entering into contracts, food processors should analyse each logistics activity in the value-creation system carefully and outsource only those activities that have low logistics asset specificity, less closeness to core business or high supply chain complexity.

Second, only outsource all logistics activities if the goal is to improve service performance

Many market survey reports show that the 'service level is not achieved' is often clustered at the top list of complains about LSPs (Capgemini, 2007; Wilding and Juriado, 2004). Food processors should bear in mind that service performance can not easily be achieved through outsourcing. Our data indicates that if improving service performance is the goal for food processors, it is necessary to outsource all logistics activities, i.e. the Level 4 to a qualified LSP that provides superior services. Thus, these LSPs can retain full control over the logistics operation and respond more quickly to changing customer needs.

To logistics service providers

First, design different service systems for different types of food products

Our results suggest that processors of chilled food require more service-driven services than non-chilled food processors. Processors of chilled food require more flexible, more reliable and quick delivery speed than non-chilled food processor. It is important that LSPs design different service systems for these two different food processors. For serving chilled food processors, in particular, LSPs could recruit employees who have knowledge about food quality and safety management, or assign more employees to handle chilled food or implement information technology to manage logistics flows of chilled food products.

Second, offer level 2 and 4 services in the Taiwanese market

Our research reveals that outsourcing value-added and distribution network design activities are likely to increase in the future in Taiwan. Although outsourcing will increase, at the moment it is hard to find local LSPs that can provide a wide range of services comparable to those in the Dutch logistics industry. Therefore, we suggest that pure cost oriented local LSPs in Taiwan review their current strategy and decide if they want to continue their strategy with low costs and low profits or migrate to become a differentiation-oriented provider in order to achieve better business performance.

6.6 Final remarks

The present research has documented the factors that determine different outsourcing levels of logistics activities in Taiwan and the Netherlands. Based on an analysis of data collected from three cases and a survey of 114 food processors, the results indicate that asset specificity, core closeness, supply chain complexity, and logistics strategy are important for logistic outsourcing decisions. The results also suggest that outsourcing decisions of different levels of logistics activities have different determining factors. In addition, outsourcing of the highest outsourcing levels, such as outsourcing of selecting logistics companies or selecting factory locations results in better service performances. Regarding the country comparison, the Netherlands has higher percentages for outsourcing of transportation (Level 1) and tactical planning activities (Level 3). When plans for the future are included, Taiwan will probably outsource value added activities (Level 2) and strategic planning activities (Level 4) much more than the Netherlands.

These findings supported the transaction cost and resource-based view theories used in this research. Although the two theories have different views of the factors that determine a firm's outsourcing decision, the analysis shows that the two are complementary. The findings also contribute to the supply chain management theory regarding supply chain characteristics by stressing that the degree of collaboration between a food processor and a logistics service provider might be dependent on the degree of supply chain complexity.

These findings provide equally interesting views for scholars on theoretical debates and for food processors who are looking for a successful outsourcing experience, and also for logistics service providers who are looking for potential customers.

References

Aktas, E. and F. Ulengin, 2005. Outsourcing logistics activities in Turkey. The Journal of Enterprise Information Management 18 (3): 316-329.

Al-kaabi, H., A. Potter and M. Naim, 2007. An outsourcing decision model for airlines' MRO activities. Journal of Quality in Maintenance Engineering 13 (3): 217-227.

Andersson, P., H. Aronsson and N.G. Storhagen, 1989. Measuring logistics performance. Engineering Costs and Production Economics 17: 253-262.

Argyres, N., 1996. Evidence on the Role of Firm Capabilities in Vertical Integration Decisions. Strategic Management Journal 17 (2): 129-150.

Arnold, U., 2000. New dimensions of outsourcing: a combination of transaction cost economics and the core competencies concept. European Journal of Purchasing & Supply Management 6: 23-29.

Arroyo, P., J. Gaytan and L.d. Boer, 2006. A survey of third party logistics in Mexico and a comparison with reports on Europe and USA. International Journal of Operation & Production Management 26 (6): 639-667.

Aubert, B.A., S. Rivard and M. Patry, 1996. A transaction cost approach to outsourcing behavior: Some empirical evidence. Information & Management 30 (2): 51-64.

Aubert, B.A., S. Rivard and M. Patry, 2004. A transaction cost model of IT outsourcing. Information & Management 41 (7): 921-932.

Bagchi, P.K. and H. Virum, 1996. European logistic alliances: a management model. International Journal of Logistics Management, 7 (1): 93-108.

Ball, D.P., 2007. Uncertainty in supply chain configuration and operation. Production Planning & Control 18 (6): 453-453.

Barney, J., 1991. Firm resources and sustained competitive advantage. Journal of Management 17 (1): 99-121.

Barney, J.B., 2007. Gaining and sustaining competitive advantage, Pearson Education, New Jersey.

Barthelemy, J., 2003. The seven deadly sins of outsourcing. Academy of Management Executive 17 (2): 87-98.

Beamon, B.M., 1999. Measuring supply chain performance. International Journal of Operation & Production Management 19 (3): 275-292.

Beaumont, N. and A. Sohal, 2004. Outsourcing in Australia. International Journal of Operation & Production Management 24 (7): 688-700.

Berglund, M., P.V. Laarhoven, G. Sharman and S. Wandel, 1999. Third-party logistics: is there a future? International Journal of Logistics Management 10 (1): 59-70.

Bhatnagar, R. and S. Viswanathan, 2000. Re-engineering global supply chain. International Journal of Physical Distribution & Logistics Management 30 (1): 13-34.

Bolumole, Y.A., 2001. The supply chain role of third-party logistics providers. International Journal of Logistics Management 12 (2): 87-102.

Bourlakis, M.A. and P.W.H. Weightman, 2004. Food supply chain management, Blackwell Publishing, Oxford.

Bowersox, D.J. and D.J. Closs, 1996. Logistics Management: the integrated supply chain processes, McGraw Hill, New York.

References

Bowler, R.A., 1975. Logistics and the failure of the British army in America 1775-1783, Princeton University Press.

Brandes, H., J. Lilliecreutz and S. Brege, 1997. Outsourcing-success or failure? European Journal of Purchasing & Supply Management 3 (2): 63-75.

Bremmers, H.J., S.W.F. Omta, J.H. Trienekens and E.F.M. Wubben, 2004. Dynamics in chains and networks: proceedings of the sixth international conference on chain and network management in agribusiness and the food industry (Ede, 27-28 May 2004), Wageningen Academic Publishers, Wageningen, 630 pp.

Bstieler, L., 2005. The moderating effect of environmental uncertainty on new product development and time efficiency. The Journal of Product Innovation Management 22: 267-284.

Carbone, V. and M.A. Stone, 2005. Growth and relational strategies used by the European logistics service providers: Rationale and outcomes. Transportation Research Part E 41: 495-510.

Choi, T.Y. and D.R. Krause, 2006. The supply base and its complexity: implications for transaction costs, risks, responsiveness, and innovation. Journal of Operation Management 24: 637-652.

Christopher, M., 2005. Logistics and supply chain management Creating value-adding networks, Pearson Education Limited, Harlow.

Coase, R.H., 1937. The nature of the firm. Economica 4 (16): 386-405.

Conner, K.R. and C.K. Prahalad, 1996. A resource-based theory of the firm: knowledge versus opportunism. Organization Science 7 (5): 477-501.

Cullen, J.B. and K.P. Parboteeah, 2005. Multinational management: a strategic approach, South-Western, United States.

Dabhilkar, M. and L. Bengtsson, 2008. Invest or divest? On the relative improvement potential in outsourcing manufacturing. Production Planning & Control 19 (3): 212-228.

Dankbaar, B., 2007. Global sourcing and innovation: The consequences of losing both organizational and geographical proximity. European Planning Studies 15 (2): 271-288.

Dapiran, P., R. Lieb and R. Millen, 1996. Third party logistics services usage by large Australian firms. International Journal of Physical Distribution & Logistics Management 26 (10): 36-45.

David, R.J. and S.K. Han, 2004. A systematic assessment of the empirical support for transaction cost economics. Strategic Management Journal 25 (1): 39-58.

Dean, J.W.J. and M.P. Sharfman, 1993. The relationship between procedural rationality and political behavior in strategic decision making. Decision Sciences 24: 1069-1083.

Drazin, R. and A.H. Van de Ven, 1985. Forms of fit in contingency theory. Administrative Science Quarterly 30 (4): 514-539.

Dyer, J.H., 1997. Effective interfirm collaboration: How firms minimize transaction costs and maximize transaction value. Strategic Management Journal 18 (7): 535-556.

Eisenhardt, K.M., 1989. Making Fast Strategic Decisions in High-Velocity Environments. The academy of Management Journal 32 (3): 543-576.

Feng, C.M. and K.C. Chia, 2000. Logistics opportunities in Asia and development in Taiwan. Transport Reviews 20 (2): 257-265.

Folkers, H. and H. Koehorst, 1997. Challenges in international food supply chains: vertical co-ordination in the European agribusiness and food industries. Supply Chain Management 2 (1): 11-14.

Forza, C., 2002. Survey research in operations management: a process-based perspective. International Journal of Operation & Production Management 22 (2): 152-194.

Franceschini, F., M. Galetto, A. Pigaatelli and M. Varetto, 2003. Outsourcing: guidelines for a structured approach. Benchmarking: An International Journal 10 (3): 246-260.

Germain, R., C. Claycomb and C. Dro¨ge, 2008. Supply chain variability, organizational structure, and performance: The moderating effect of demand unpredictability. Journal of Operations Management 26: 557-570.

Gilley, K.M., C.R. Greer and A.A. Rasheed, 2004. Human resource outsourcing and organizational performance in manufacturing firms. Journal of Business Research 57: 232-240.

Gilley, K.M., J.E. McGee and A.A. Rasheed, 2004. Perceived Environmental Dynamism and Managerial Risk Aversion as Antecedents of Manufacturing Outsourcing: The Moderating Effects of Firm Maturity. Journal of Small Business Management 42 (2): 117-133.

Gilley, K.M. and A. Rasheed, 2000. Making more by doing less: An analysis of outsourcing and its effects on firm performance. Journal of Management 26 (4): 763-790.

Grievink, J.W., L. Josten and C. Valk, 2002. State of the art in food: the changing face of the worldwide food industry, Elsevier Business Information, Meppel.

Grover, V. and M.K. Malhotra, 2003. Transaction cost framework in operations and supply chain management research: theory and measurement. Journal of Operations Management 21 (4): 457-473.

Groves, R.M., F.J. Fowler, M.P. Couper, J.M. Lepkowski, E. Singer and R. Tourangeau, 2004. Survey methodology, Wiley-Interscience, New York.

Guimaraes, T., N. Martensson, J. Stahre and M. Igbaria, 1999. Empirically testing the impact of manufacturing system complexity on performance. International Journal of Operation & Production Management 19 (12): 1254-1269.

Hafeez, K., Y.B. Zhang and N. Malak, 2002. Core competence for sustainable competitive advantage: A structured methodology for identifying core competence. Ieee Transactions on Engineering Management 49 (1): 28-35.

Hair, J.F., R.E. Anderson, R.L. Tatham and W.C. Black, 1998. Multivariate data analysis, Prentice-Hall, New Jersey.

Halldorsson, A. and T. Skjott-Larsen, 2004. Developing logistics competencies through third party logistics relationships. International Journal of Operation & Production Management 24 (2): 192-206.

Hertz, S. and M. Alfredsson, 2003. Strategic development of the third party logistics providers. Industrial Marketing Management 32: 139-179.

Holcomb, T.R. and M.A. Hitt, 2007. Toward a model of strategic outsourcing. Journal of Operations Management 25 (2): 464-481.

Hong, J., A.T.H. Chin and B. Liu, 2004. Logistics outsourcing by manufacturers in china: a survey of the industry. Transportation Journal Winter: 17-25.

Hosmer, D.W. and S. Lemeshow, 1989. Applied logistic regression, John Wley & Sons, Inc., USA.

Hough, J.R. and M.A. White, 2003. Environmental dynamism and strategic decision making rationality: an examination at the decision level. Strategic Management Journal 24: 481-489.

Hsiao, H.I., R.G.M. Kemp, J.G.A.J. van der Vorst and S.W.F. Omta, 2008. Make-or-buy decisions in logistics: an empirical analysis. Electronic proceedings of the 8th International Conference on Management in AgriFood Chains and Networks 28-30 May 2008, Ede, the Netherlands.

Hsiao, H.I., J.G.A.J. van der Vorst and S.W.F. Omta, 2006. Logistics Outsourcing in Food Supply Chain Networks. In: Bijman, J., S.W.F. Omta, J.H. Trienekens, J.H.M. Wijnands and E.F.M. Wubben (eds.), International agri-food chains and networks: management and organization Wageningen Academic Publishers, Wageningen, the Netherlands, pp. 135-150.

Insinga, R.C. and M.J. Werle, 2000. Linking outsourcing to business strategy. The Academy of Management Executive 14 (4): 58-70.

Jaccard, J. and R. Turrisi, 2003. Interaction effects in multiple regression, Sage Publications.

Jacobides, M.G. and L.M. Hitt, 2005. Losing sight of the forest for the trees? Productive capabilities and gains from trade as drivers of vertical scope. Strategic Management Journal 26 (13): 1209-1227.

Jeng, H.Y.J. and T.J. Fang, 2003. Food safety control system in Taiwan-the example of food service sector. Food Control 14: 317-322.

Jennings, D., 1997. Strategic guildlines for outsourcing decisions. Strategic Change 6: 85-96.

Jiang, B., J.A. Belohlav and S.T. Young, 2007. Outsourcing impact on manufacturing firms's value: Evidence from Japan. Journal of Operation Management 25: 885-900.

Jiang, B., G.V. Frazier and E.L. Prater, 2006. Outsourcing effects on firms' operational performance an empirical study. International Journal of Operation & Production Management 26 (12): 1280-1300.

Judge, W.Q. and A. Miller, 1991. Antecedents and Outcomes of Decision Speed in Different Environmental Contexts. The academy of Management Journal 34 (2): 449-463.

Kisperska-Moron, D., 2005. Logistics customer service levels in Poland: Changes between 1993 and 2001. International Journal of Production Economics 93-94: 121-128.

Lai, K.H., 2004. Service capability and performance of logistics service providers. Transportation Research Part E 40: 385-399.

Larson, P.D. and J.D. Kulchitsky, 1999. Logistics improvement programs The dynamics between people and performance. International Journal of Physical Distribution & Logistics Management 29 (2): 88-102.

Lau, K.H. and J. Zhang, 2006. Drivers and obstacles of outsourcing practices in China. International Journal of Physical Distribution & Logistics Management 36 (10): 776-792.

Lawrence, P.R. and J.W. Lorsch, 1967. Organization and Environment Managing Differentiation and Integration, Division of Research Graduate School of Business Administration Harvard University, Boston.

Leiblein, M.J. and D.J. Miller, 2003. An empirical examination of transaction-and firm-level influences on the vertical boundaries of the firm. Strategic Management Journal 24: 839-859.

Londe, B.L. and M. Cooper, 1998. Partnership in providing customer service: a third-party perspective, Oak Brook.

Madhok, A., 2002. Reassessing the fundamentals and beyond: Ronald Coase: the transaction cost and resource-based theories of the firm and the institutional structure of production. Strategic Management Journal 23: 535-550.

Mapes, J., M. Szwejczewski and C. New, 2000. Process variability and its effect on plant performance. International Journal of Operations & Production Management 20 (7): 792-808.

McAuley, J., J. Duberley and P. Johnson, 2007. Organization Theory, Pearson Education Limited.

McIvor, R.T., P.K. Humphrey and W.E. Mcaleer, 1997. A strategic model for the formulation of an effective make or buy decision. Management Decision 35 (2): 169-178.

Milgate, M., 2001. Supply chain complexity and delivery performance: an international exploratory study. Supply chain management: An international Journal 6 (3): 106-118.

Millen, R., A. Sohal, P. Dapiran, R. Lieb and L.N.V. Wassenhove, 1997. Benchmarking Australian firms' usage of contract logistics services: a comparison with American and Western European practice Benchmarking for Quality Management & Technology 4 (1): 34-46.

Norek, C.D. and T.L. Pohlen, 2001. Cost knowledge: a foundation for improving supply chain relationships. International Journal of Logistics Management 12 (1): 37-51.

Oh, J. and S.K. Rhee, 2008. The influence of supplier capabilities and technology uncertainty on manufacturer-supplier collaboration A study of the Korean automotive industry. International Journal of Operation & Production Management 28 (6): 490-517.

Olavarrieta, S. and A.E. Ellinger, 1997. Resource-based theory and strategic logistics research International Journal of Physical Distribution & Logistics Management 29 (910): 559-587.

Omta, S., 2004. Increasing the innovative potential in chains and networks. Journal on Chain and Network Science 4 (2): 75-81.

Pache, G., 1998. Logistics outsourcing in grocery distribution:a European perspective. Logistics Information Management 11 (5): 301-308.

Perrow, C., 1967. A Framework for the Comparative Analysis of Organizations. American Sociological Review 32 (4): 194-208.

Poppo, L. and T. Zenger, 1998. Testing alternative theories of the firm: Transaction cost, knowledge-based, and measurement explanations for make-or-buy decisions in information services. Strategic Management Journal 19 (9): 853-877.

Power, D., M. Sharafali and V. Bhakoo, 2006. Adding value through outsourcing Contribution of 3PL services to customer performance. Management Research News 30 (3): 228-235.

Prahalad, C.K. and G. Hamel, 1990. The core competence of the corporation Harvard Business Review (May-June): 79-91.

Quinn, J.B. and F.G. Hilmer, 1994. Strategic outsourcing. Sloan Management Review: 43-55.

Rao and Young, 1994. Global supply chains: factors influencing outsourcing of logistics functions. International Journal of Physical Distribution & Logistics Management 24 (6): 11-19.

Razzaque, M.A. and C.C. Sheng, 1998. Outsourcing of logistics functions: a literature survey. International Journal of Physical Distribution & Logistics Management 28 (3): 89-107.

Rindfleisch, A. and J.B. Heide, 1997. Transaction cost analysis: past, present and future applications. Journal of Marketing 61: 30-54.

Robertson, T.S. and H. Gatignon, 1998. Technology development mode: a transaction cost conceptualization Strategic Management Journal 19: 515-531.

Safizadeh, M.H., J.M. Field and L.P. Ritzman, 2008. Sourcing practices and boundaries of the firm in the financial services industry. Strategic Management Journal 29: 79-91.

Sahay, B.S. and R. Mohan, 2006. 3PL practices: an Indian perspective. International Journal of Physical Distribution & Logistics Management 36 (9): 666-689.

Salimath, M.S., J.B. Cullen and U.N. Umesh, 2008. Outsourcing and performance in entrepreneurial firms: contingent relationships with entrepreneurial configurations. Decision Sciences 39 (3): 359-381.

Sink, H.L. and C.J. Langley, 1997. A managerial framework for the acquisition of third-party logistics services. Journal of Business Logistics 18 (2): 163-189.

Sohail, M.S., R. Bhatnagar and A.S. Sohal, 2006. A comparative study on the use of third party logistics services by Singaporean and Malaysian firms. International Journal of Physical Distribution & Logistics Management 36 (9): 690-701.

Stadtler, H., 2002. Supply chain management and advanced planning, Springer-Verlag Heidelberg.

Stank, T.P., T.J. Goldsby, S.K. Vickery and K. Savitskie, 2003. Logistics service performance: estimating its influence on market share. Journal of Business Logistics 24 (1): 27-55.

Stevens, G.C., 1989. Integrating the supply chain. International Journal of Physical Distribution & Material Management 19 (8): 3-8.

Sum, C.C. and C.B. Teo, 1999. Strategic posture of logistics service providers in Singapore. International Journal of Physical Distribution & Logistics Management 29 (9): 588-605.

Taylor, D.H., 2006. Strategic considerations in the development of lean agri-food supply chains: a case study of the UK pork sector. Supply Chain Management-an International Journal 11 (3): 271-280.

Teng, J.T.C., M.J. Cheon and V. Grover, 1995. Decisions to Outsource Information-Systems Functions - Testing a Strategy-Theoretic Discrepancy Model. Decision Sciences 26 (1): 75-103.

Traill, W.B. and E. Pitts, 1998. Competitiveness in the food industry, Blackie Academic & Professional, London.

Van Damme, D.E. and M.J.P. Van Amstel, 1996. Outsourcing logistics management activities. International Journal of Logistics Management 7 (2): 85-95.

Van der Vorst, J.G.A.J., 2000. Effective food supply chain-generating, modeling and evaluating supply chain scenarios, PhD thesis, Wageningen University, The Netherlands.

Van der Vorst, J.G.A.J. and A.J.M. Beulens, 2002. Identifying sources of uncertainty to generate supply chain redesign strategies. International Journal of Physical Distribution & Logistics Management 32 (6): 409-430.

Van der Vorst, J.G.A.J., A.J.M. Beulens and P.V. Beek, 2005. Innovations in logistics and ICT in food supply chain networks In: Jongen, W.M.F. and M.T.G. Meulenberg (eds.), Innovation in Agri-Food Systems, Wageningen Adademic Publishers, Wageningen, the Netherlands, pp. 245-292.

Van der Vorst, J.G.A.J., M.P.J. Duineveld, F.P. Scheer and A.J.M. Beulens, 2007. Towards logistics orchestration in the pot plant supply chain network. Electronic proceedings of the Euroma 2007 conference. 18-20 June 2007, Ankara, Turkey, pp. 1-10.

Van Donk, D.P. and T. Van der Vaart, 2005. A case of shared resources,uncertainty and supply chain integration in the process industry. International Journal of Production Economics 96: 97-108.

Van Duren, E. and D. Sparling, 1998. Supply chain management and the Canadian agri-food sector. Canadian Journal of Agricultural Economics-Revue Canadienne D Agroeconomie 46 (4): 479-489.

Van Goor, A.R., M.J.P.V. Amstel and W.P.V. Amstel, 2003. European distribution and supply chain logistics, Wolters-Noordhoff, Groningen.

Voss, C., N. Tsikriktsis and M. Frohlich, 2002. Case research in operations management. International Journal of Operation & Production Management 22 (2): 195-219.

Waldman, D.A., G.G. Ramirez, R.J. House and P. Puranam, 2001. Does leadership matter? CEO leadership attributes and profitability under conditions of perceived environmental uncertaitny Academy of Management Journal 44 (1): 134-143.

Wang, Q. and N. Von Tunzelmann, 2000. Complexity and the functions of the firm: breadth and depth. Research Policy 29 (7-8): 805-818.

Wanke, P.F. and W. Zinn, 2004. Strategic logistics decision making. International Journal of Physical Distribution & Logistics Management 34 (6): 466-478.

Wernerfelt, B., 1984. A resource-based view of the firm. Strategic Management Journal 5 (2): 171-180.

Westgren, R.E., 1998. Innovation and future directions of supply chain management in US agri-food. Canadian Journal of Agricultural Economics-Revue Canadienne D Agroeconomie 46 (4): 519-524.

Wheelwright, S.C., 1984. Manufacturing strategy:defining the missing link. Strategic Management Journal 5 (1): 77-91.

Wilding, R. and R. Juriado, 2004. Customer perceptions on logistics outsourcing in the European consumer goods industry. International Journal of Physical Distribution & Logistics Management 34 (8): 628-644.

Williamson, O.E., 1975. Markets and Hierarchies: Analysis and Antitrust Implications, Free Press, New York.

Williamson, O.E., 1985. The economic Institutions of Capitalism, Free Press, New York.

Williamson, O.E., 1991. Comparative economic organization: the analysis of discrete structural alternatives. Administrative Science Quarterly 36: 269-296.

Williamson, O.E., 1998. Transaction cost economics; how it works; where it is headed. De Economist 146 (1): 23-58.

Wong, C.Y. and S. Boon-itt, 2008. The influence of institutional norms and environmental uncertainty on supply chain integration in the Thai automotive industry. International Journal of Production Economics 115: 400-410.

Yasuda, H., 2005. Formation of strategic alliances in high-technology industries: comparative study of the resource-based theory and the transaction-cost theory. Technovation 25: 763-770.

Websites

Capgemini, 2001. Third-party logistics study. Available at: http://bus.utk.edu/ivc/supplychain/Readings/SC_B2B_3PLogistics.pdf. Accessed 10 Feb 2003.

Capgemini, 2002. Third-party logistics study. Available at: http://www.at.cgey.com/servlet/PB/show/1005207/3PL_Study.pdf. Accessed Sep 2003.

Capgemini, 2003. Third-party logistics. Available at: http://www.capgemini.com/. Accessed 15 Nov 2005.

Capgemini, 2005. Third-party logistics study. Available at: http://www.at.cgey.com. Accessed 5 Sep 2006.

Capgemini, 2006. Europe's most wanted distribution center locations. Available at: www.capgemini.com. Accessed 2 Sep 2008.

References

Capgemini, 2007. Third-party logistics. Available at: http://www.at.cgey.com. Accessed 11 Oct 2008.

CEPD (Council For Economic Planning And Development), 2002. Challenge 2008- National Development Plan. Available at: www.cepd.gov.tw. Accessed 1 Sep 2008.

Lieb, R., 2002. The use of third party logistics services by large American manufacturers, the 2002 survey. Available at: http://www.accenture.com/xdoc/en/services/scm/scm_thought_user_survey.pdf. Accessed Dec 2004.

Port of Rotterdam, 2007. Port statistics. available at: www.portofrotterdam.com (accessed 11 Sep 2008).

WTO (World Trade Organization), 2008. International Trade Statistics 2008. available at: www.wto.org/english/res_e/statis_e/its2008_e/its08_toc_e.htm (accessed 11 Feb 2009).

Appendices

Appendix 1. Survey measurements

1. *Make-or-buy choices* The outsourcing status was assessed using a two-point scales with two anchors (have outsourced, and won't outsource).
2. *Asset specificity* To measure this variable, we use 'We have invested in special equipments to conduct this activity' (Poppo and Zenger, 1998). This item is measured using 10-point scales anchored by 'strongly disagree' and 'strongly agree.'
3. *Performance measurement uncertainty* To measure this variable, we use 'We specify precise measures for evaluating the performance of this activity' (Robertson and Gatignon, 1998). This item is measured using 10-point scales anchored by 'strongly disagree' and 'strongly agree.'
4. *Core closeness* To measure this variable, we use 'This activity contributes highly to our competitive advantage.' This item is measured using 10-point scales anchored by 'strongly disagree' and 'strongly agree.'
5. *Supply chain complexity* A four measurement items are obtained from cases results to measure this variable. These items are 'number of stock keeping units,' 'demand uncertainty,' 'number of international customers,' and 'distribution channel variety' Respondents were asked to rank 1 to 7 scales indicating to what extent they agree if these factors complicate the management of logistics processes (1=strongly disagree; 7=strongly agree).
6. *Logistics strategy* We measure this variable using two scales anchored by cost, flexibility and food quality. Respondents were asked to rank each of objectives its importance in percentage with overall sum of 100.

Appendix 2. Measures and Cronbach alphas

Asset specificity ($\alpha = 0.69$)
- We have invested in special equipments to conduct this activity
- We have acquired special knowledge and skills to perform this activity
- It is very costly to outsource this activity

Performance measurement uncertainty
- It is difficult to measure the performance of logistics service providers for this activity

Core business closeness ($\alpha = 0.75$)
- This activity contributes highly to our competitive advantage
- This activity is essential to support our core activities
- Compared to our rivals, this activity is performed efficiently

Supply chain complexity

Distribution complexity ($\alpha = 0.82$)
- Number of packaging lines
- Number of clients
- Delivery frequency
- Lead time requirement

Distribution network complexity ($\alpha = 079$)
- Storage variety
- Number of warehouses
- Distribution channel variety
- Distribution uncertainty

Demand complexity ($\alpha = 0.83$)
- Demand volume
- Demand uncertainty
- Demand fluctuation

Firm size
- Full-time employees

Changes of sales growth rate
- Development of total sales volume over the 2003-2005
- Expected development of total sales volume over the 2005-2008

Appendix 3. Descriptive statistics and correlation

Variable	Mean	S.D.	1	2	3	4	5	6	7	8	9
Transportation											
1. Outsourcing decision	0.69	0.463									
2. Location	1.39	0.491	-0.163								
3. Size	3.7632	1.45920	0.114	0.515**							
4. Sale growth	1.61	0.763	-0.057	0.006	-0.085						
5. Asset specificity	4.9181	2.15549	-0.386**	0.153	0.134	-0.145					
6. Performance measurement uncertainty	5.9649	2.43474	0.116	-0.462**	-0.241**	-0.068	0.034				
7. Core closeness	7.07164	1.800520	0.037	0.245*	0.233*	-0.160	0.394**	-0.057			
8. Distribution complexity	4.1091	1.44664	-0.063	-0.116	0.034	-0.024	0.129	0.128	-0.059		
9. Distribution channel complexity	3.4299	1.54249	0.111	0.036	0.204*	-0.078	-0.152	0.046	-0.047	0.594**	
10. Demand complexity	4.5929	1.56013	0.128	-0.028	0.108	-0.065	-0.137	-0.009	-0.082	0.432**	0.436**
Packaging											
1. Outsourcing decision	0.16	0.366									
2. Location	1.39	0.491	0.093								
3. Size	3.7632	1.45920	0.253**	0.515**							
4. Sale growth	1.61	0.763	-0.043	0.006	-0.085						
5. Asset specificity	6.2105	2.10422	-0.208*	-0.121	0.012	-0.027					
6. Performance measurement uncertainty	4.7321	2.74759	0.052	-0.086	-0.048	-0.079	0.189*				
7. Core closeness	6.8143	2.11909	0.034	-0.025	0.166	-0.084	0.666**	0.164			
8. Distribution complexity	4.1091	1.44664	0.194*	-0.116	0.034	-0.024	-0.002	0.138	-0.087		
9. Distribution channel complexity	3.4299	1.54249	0.185*	0.036	0.204*	-0.078	0.076	0.171	0.042	0.594**	
10. Demand complexity	4.5929	1.56013	0.099	-0.028	0.108	-0.065	-0.051	0.005	-0.145	0.432**	0.436**

N=114; *$P<0.05$; ** $P<0.01$ (2-tailed).

Variable	Mean	S.D.	1	2	3	4	5	6	7	8	9
Transportation management											
1. Outsourcing decision	0.37	0.485									
2. Location	1.39	0.491	-0.096								
3. Size	3.7632	1.45920	0.037	0.515**							
4. Sale growth	1.61	0.763	-0.049	0.006	-0.085						
5. Asset specificity	5.3494	2.17137	-0.314**	0.130	0.228*	0.007					
6. Performance measurement uncertainty	5.3243	2.43893	0.087	-0.283**	-0.071	-0.032	0.116				
7. Core closeness	5.2865	1.52648	-0.347**	0.163	0.219*	0.009	0.592**	0.045			
8. Distribution complexity	4.1091	1.44664	0.051	-0.116	0.034	-0.024	0.098	0.212*	-0.001		
9. Distribution channel complexity	3.4299	1.54249	0.055	0.036	0.204*	-0.078	0.002	0.120	0.044	0.594**	
10. Demand complexity	4.5929	1.56013	0.107	-0.028	0.108	-0.065	-0.074	0.083	-0.154	0.432**	0.436**
Distribution network design											
1. Outsourcing decision	0.11	0.308									
2. Location	1.39	0.491	0.015								
3. Size	3.7632	1.45920	-0.023	0.515**							
4. Sale growth	1.61	0.763	-0.023	0.006	-0.085						
5. Asset specificity	5.3799	2.29703	-0.109	0.067	0.309**	0.045					
6. Performance measurement	4.5185	2.50413	-0.014	-0.040	0.060	0.091	0.408**				
7. Core closeness	6.3363	2.29396	-0.038	-0.041	0.259**	-0.019	0.641**	0.206*			
8. Distribution complexity	4.1091	1.44664	0.004	-0.116	0.034	-0.024	0.116	-0.005	0.026		
9. Distribution channel complexity	3.4299	1.54249	-0.012	0.036	0.204*	-0.078	0.064	-0.176	0.063	0.594**	
10. Demand complexity	4.5929	1.56013	0.170	-0.028	0.108	-0.065	0.051	-0.077	0.036	0.432**	0.436**

N=114; *$P<0.05$; ** $P<0.01$ (2-tailed).

Appendix 4 Questionnaire

Part I: General information

1. What is your job position?
☐ Logistics manager   ☐ Financial manager   ☐ Production manager   ☐ Director
☐ Other_____

2. In your company, at what level the decision of logistics outsourcing is taken?
☐ one factory level   ☐ multi-factories level

3. To what degree, are you involved in decision making of logistics outsourcing?
☐ highly   ☐ moderately   ☐ a little bit   ☐ not involved

4. How many full-time employees are there at your company?
☐ Less than 40   ☐ 40- <50   ☐ 50-<100   ☐ 100-<150
☐ 150-<250   ☐ larger than 250

5. What was the development of your total sales volume over the next three years?
2006-2008 _____% (increase or decrease in %)
and the expected development of your total sales volume over the last three year?
2003-2005_____%(increase or decrease in %)

6. How many product groups does your company have in the Netherlands?
☐ I   ☐ 2   ☐ 3   ☐ 4   ☐ >4

Please choose the product group with the largest sales volume in the last year to answer following questions:
*Please first identify this product group (e.g. yogurt, veal, soup): _____

How large is the sales percentage of this product group to your company? *(check one of boxes)*
☐ <20%   ☐ 20%-40%   ☐ 40%-60%   ☐ 60%-80%   ☐ 80%-100%   ☐ 100%

Please describe storage characteristics of this product group? *(multiple choices are possible)*
☐ frozen   ☐ chilled   ☐ room temperature   ☐ humidity controlled
☐ atmosphere controlled

Part 2: Outsourcing of logistics activities in your product group

We would like to obtain the insights of your current practice of logistics outsourcing in your product group, and we would like to know the reasons why you outsource some of the logistics activities.

Strongly disagree							**Strongly agree**		
1	2	3	4	5	6	7	8	9	10

	Execution level		Planning level		Strategic planning level
	Outbound transportation	Packaging & labelling	Transportation management	Inventory management	Distribution network planning
a. We have already outsourced this activity	☐ a	☐ a	☐ a	☐ a	☐ a
b. We intend to outsource this activity	☐ b	☐ b	☐ b	☐ b	☐ b
c. We don't intend to outsource this activity	☐ c	☐ c	☐ c	☐ c	☐ c

1. This activity contributes highly to our competitive advantage.
2. This activity is essential to support our core activities.
3. Compared to our rivals, this activity is performed efficiently.
4. We have invested in special equipments to conduct this activity.
5. We have acquired special knowledge and skills to perform this activity.
6. It is very costly to outsource this activity.
7. We specify precise measures for evaluating the performance of this activity.
8. It is difficult to measure the performance of logistics service providers for this activity.

Part 3: Logistics complexity in your product group

We would like to know to what extent you agree if the following factors complicate the management of logistics processes in your product group.

Strongly disagree						**Strongly agree**	
I	2	3	4	5	6		7

End product characteristics

Perishability of end products	I	2	3	4	5	6	7
Number of products (SKUs)	I	2	3	4	5	6	7
Number of product groups	I	2	3	4	5	6	7
Variety of product in storage conditions	I	2	3	4	5	6	7

Production characteristics

Number of packaging lines	I	2	3	4	5	6	7
Uncertainty of production output time, quantity and quality	I	2	3	4	5	6	7

Sales/Demand characteristics

Annual demand volume	I	2	3	4	5	6	7
Demand uncertainty	I	2	3	4	5	6	7
Demand fluctuation	I	2	3	4	5	6	7

Distribution characteristics

Number of customers	I	2	3	4	5	6	7
Number of international customers	I	2	3	4	5	6	7
Number of warehouses	I	2	3	4	5	6	7
Distribution channel variety	I	2	3	4	5	6	7
Delivery frequency	I	2	3	4	5	6	7
Order lead time	I	2	3	4	5	6	7
Distribution batch size	I	2	3	4	5	6	7
Uncertainty of distribution time, quantity and quality	I	2	3	4	5	6	7

Part 4: Your logistics strategy

Please indicate at a scale of 0-100 points the relative importance of the following objectives

Logistics objectives	Points
Low logistics cost	
High reliable and consistent logistics service	
Short delivery lead time	
High flexibility to accommodate demand changes	
High Food quality	
	Total 100

If there are other objectives, please specify:

Part 5: Your logistics performance

Please provide the following information with respect to your current logistics performance.

Strongly disagree					Strongly agree	
I	2	3	4	5	6	7

Comparing with our competitors...

Our logistics costs are relatively low	I	2	3	4	5	6	7
We always meet the promised delivery time	I	2	3	4	5	6	7
We always meet the ordered quantity	I	2	3	4	5	6	7
We quickly respond to the needs of our key customers	I	2	3	4	5	6	7
We offer shorter lead-time	I	2	3	4	5	6	7
We relatively offer longer shelf life	I	2	3	4	5	6	7

Summary

Our research interest began by observing some dramatic changes in the food processing industry and noticing some new services emerging in the logistics industry. As the competitive environment of food business continues to change, traditional methods of managing logistics flows might no longer be valid for ensuring the firm's performance. In recent years, there has been an increased academic interest and a large number of scientific publications in the area of logistics outsourcing (Aktas and Ulengin, 2005; Berglund *et al.*, 1999; Bolumole, 2001; Hertz and Alfredsson, 2003; Hong *et al.*, 2004; Lau and Zhang, 2006; Pache, 1998; Sink and Langley, 1997; Wilding and Juriado, 2004). As many industries reconfigure their operations around core competencies through outsourcing, we might wonder whether food processors will also re-examine their firm's positions within the supply chain to collaborate with third-party service providers by outsourcing some or all of their logistics activities. Outsourcing of logistics activities usually consists of four steps: (1) identification of the need, (2) selection of service providers, (3) implementation and (4) service assessment (Sink and Langley, 1997). Outsourcing can be a painful learning experience for companies because they often don't renew their contract with LSPs either because the goals are not realised or because the required service level is not achieved (Wilding and Juriado, 2004). Given these issues, literatures provide almost no guidelines identifying the most suitable logistics activities for food processing firms to outsource. To address this key problem, the objective of this book is the following:

> *to analyse how food processors determine their logistics outsourcing need and to analyse how logistics outsourcing influences logistics performance.*

This research is based on data from Dutch and Taiwanese companies. There are three reasons for this research setup. First, Taiwan is trying to become an international logistics and distribution hub in Asia-Pacific region. The Netherlands is known internationally as the logistics and distribution hub of Europe. Second, both countries are comparable in the sense that they have limited natural resources and land. Third, Dutch agriculture and food processing is famous worldwide, and agriculture is also an important industry for Taiwan. Therefore, it is interesting to compare the two countries where it is expected that Taiwan can learn from the Dutch examples. In order to realise the objective, four research questions were formulated.

1. What kind of logistics activities can be outsourced by food processors? (1a) and what decision criteria are considered when outsourcing logistics activities? (1b)
2. What decision-making criteria are considered by food processors when outsourcing a certain level of logistics activities?
3. What is the impact of logistics outsourcing on service performance?
4. What are the current and expected future developments in logistics outsourcing in Taiwan and the Netherlands?

To answer these questions, we used both case study (qualitative) and survey (quantitative) approaches. The first research question was primarily concerned with building the decision-making framework. The framework was developed on the basis of transaction cost, resource-based and supply chain management theories. The exploratory case studies were undertaken in the Netherlands to verify the factors and, possibly, identify other relevant factors that were not mentioned in the literature. The second research question was set to test the decision-making framework using Dutch (NL) and Taiwanese (TW) data. Surveys were mailed to logistics managers in food processors with at least forty employees. Data were gathered from September 2006 to February 2007. Of the 890 surveys mailed (NL: 385; TW: 505), 66 had incorrect contact information (NL: 57; TW: 9) and were returned by the postal service. A total of 138 responses were received (NL: 76; TW: 62), of which 24 had missing data (NL: 7; TW: 17) and were judged unusable, thus yielding a sample size of 114 (NL: 69; TW: 45) with a response rate of 15% (114/800) (NL: 21%; TW: 9%). The third research question was concerned with the impact of logistics outsourcing on logistics performance. We used the survey to seek to understand the impact of outsourcing decision on perceived service performance and also assessed the moderating role that supply chain complexity might play in the outsourcing-performance relationship. Finally, the fourth research question was set to investigate the current status and future development of logistics outsourcing in the Netherlands and Taiwan. Next we summarise our main results.

Regarding the first research question, *what kind of logistics activities can be outsourced by food processors? (1a) and what decision criteria are considered when outsourcing logistics activities? (1b)* our study divided the outsourceable activities of a logistics process into four levels: level 1 refers to the execution level of basic activities, such as transportation and warehousing. Level 2 refers to value-added activities. Level 3 refers to the planning and control level, such as transportation management or inventory management and level 4 refers to strategic planning and control level, such as finding logistics companies or suggesting a factory location. Furthermore, literature and case studies suggested five factors related with outsourcing decisions. They are:
- Asset specificity refers to logistics-asset specificity. Logistics-specific assets involve investments in human or physical capital which will lose value if they are redeployed to other areas.
- Performance measurement uncertainty refers to the degree of difficulty associated with assessing the performance of transaction partners (the logistics service providers).
- Core (business) closeness refers to logistics capabilities, skills and/or experiences, with which a firm could gain greater value than competitors.
- Supply chain complexity refers to the number of elements within the focal company's logistical flow (on the bases of production, distribution and demand), and the degree to which these bases are differentiated or varied. It influences the requirements for managing the firm's logistical system.
- Logistics strategy includes three dimensions: (1) cost, (2) flexibility, and (3) food quality. Low-cost strategy refers to a strategy in which companies seek to design logistics system

more cost-efficiently than its competitors. Flexibility strategy refers to a strategy in which companies aim at being flexible to the changing and diverse needs of customers. Food quality strategy refers to a strategy in which companies aim at providing freshness, minimal damage and high food quality of a food product.

Regarding the second research question, *what decision-making criteria are considered by food processors when outsourcing a certain level of logistics activity?* our study supported the propositions that asset specificity, core closeness and supply chain complexity are related to logistics outsourcing decisions. Each level of logistics activity has its own key outsourcing determinants. In Level 1 activities, asset specificity, core closeness and low-cost strategy are important decisive factors for a food processor. In Level 2 activities, asset specificity, core closeness and the distribution complexity (characterised by a number of packaging lines, number of clients, delivery frequency and lead-time) are important decisive factors. In Level 3 activities, asset specificity, core closeness and flexibility strategy are important decisive factors for a food processor. In Level 4 activities, demand complexity (characterised by demand volume, demand uncertainty and demand fluctuation) is the decisive factor.

Regarding the third research question, *what is the impact of logistics outsourcing on service performance?* our study did not support the assumption that outsourcing has a direct impact on service performance in any of the four levels. However, we found that the relationship between outsourcing and service performance is moderated by demand complexity at level 4. This indicates that service performance of outsourcing at level 4 increases with an increasing degree of demand complexity.

Regarding the last research question, *what are the current and expected future developments in logistics outsourcing in the Netherlands and Taiwan?* we found the following. About 69% of the companies in both countries outsource Level 1 activities, 16% Level 2 and 37% Level 3 activities. Only a few companies (about 10%) outsource the highest level of activities. In particular, the Netherlands has higher percentages for Levels 1 and 3. When plans for the future are included, Taiwan will outsource Level 2 (40%) and 4 activities (36%) much more than the Netherlands (resp. 13% and 17%). When zooming in, we found that outsourcing strategies of companies in the subsectors differ. For instance, the dairy sector outsources more frequently than the meat sector on the first three levels.

Our main contribution to literature is that it has integrated different theories in one comprehensive decision-making framework. This research has put extensive effort into developing a conceptual framework that integrates insights from transaction cost theory, resource-based theory and supply chain management in one overall model. We have strong indications, both theoretical and based on empirical evidence, that such an integrative model is needed to grasp the complexity of the outsourcing decision. Furthermore, we analysed the outsourcing decision for different levels of logistics activities. Our findings indicate that each

Summary

activity has its own outsourcing considerations, thus, an analysis of the outsourcing decision for each level of logistics activities is necessary.

The implications of our study for food processors and logistics service providers are the following. We advise food processors to select the logistics activity carefully because there is no single rule for evaluating a logistics outsourcing decision for all levels of activities. Some activities consider asset specificity, and also capabilities comparing to competitors and LSPs, and some activities also consider the supply chain complexity. We also advise food processors to outsource the highest level of logistics activities if the goal is to improve service performance. We advise logistics service providers to design different service systems for different types of food products because the processors of chilled food require more flexible, more reliable and quicker delivery than the non-chilled food processors. We also advise logistics service providers to offer the value-added and distribution network design services to the Taiwanese market, because the outsourcing of these activities will potentially increase in the future in Taiwan. We suggest that pure cost oriented local LSPs in Taiwan review their current strategy and decide if they want to continue this strategy with low costs and low profits or migrate to become a differentiation-oriented provider in order to achieve better business performance.

About the author

Hsin-I Hsiao (Lilly) was born in Changhwa, Taiwan on 29 June 1976. In 2000 she received her MSc in food science at National Taiwan Ocean University with a major in food microbiology. Her MSc thesis was about prevention of fish spoilage using food microorganisms. Between the year 2000 and 2002, she worked as a marketing specialist at Jemfond Corporation (Taiwan), which is a food processor. She also worked as a supervisor of HACCP systems at Golden Dragon Lunchbox Company (Taiwan), which supplies ready-to-eat food for elementary schools. In 2002, she continued her research and received another MSc in maritime economics and logistics at the Erasmus University Rotterdam, the Netherlands. Her thesis was about distribution channel and electronic commerce in Taiwan, South Korea, and Japan. In 2003, she started working as a PhD student in the Business Administration Group of the University of Wageningen. Her current research focuses on supply chain management and innovation in the agrifood industry.

International chains and networks series

Agri-food chains and networks are swiftly moving toward globally interconnected systems with a large variety of complex relationships. This is changing the way food is brought to the market. Currently, even fresh produce can be shipped from halfway around the world at competitive prices. Unfortunately, accompanying diseases and pollution can spread equally rapidly. This requires constant monitoring and immediate responsiveness. In recent years tracking and tracing has therefore become vital in international agri-food chains and networks. This means that integrated production, logistics, information- and innovation systems are needed. To achieve these integrated global supply chains, strategic and cultural alignment, trust and compliance to national and international regulations have become key issues. In this series the following questions are answered: How should chains and networks be designed to effectively respond to the fast globalization of the business environment? And more specificly, How should firms in fast changing transition economies (such as Eastern European and developing countries) be integrated into international food chains and networks?

About the editor

Onno Omta is chaired professor in Business Administration at Wageningen University and Research Centre, the Netherlands. He received an MSc in Biochemistry and a PhD in innovation management, both from the University of Groningen. He is the Editor-in-Chief of The Journal on Chain and Network Science, and he has published numerous articles in leading scientific journals in the field of chains and networks and innovation. He has worked as a consultant and researcher for a large variety of (multinational) technology-based prospector companies within the agri-food industry (e.g. Unilever, VION, Bonduelle, Campina, Friesland Foods, FloraHolland) and in other industries (e.g. SKF, Airbus, Erickson, Exxon, Hilti and Philips).

Guest editors

Ron Kemp is a Senior Researcher at the Bureau of the Chief Economist of the Netherlands Competition Authority and an Assistant Professor at Wageningen University and Research Centre, The Netherlands. His research focuses on competition issues, growth of SMEs and innovation.

Jack G.A.J. van der Vorst (1970) is Full Professor of Logistics and Operations Research and head of this group at Wageningen University in the Netherlands. He obtained his Masters degree 'Agro-Logistics Management' with honour in 1994 and his Phd degree in 2000 with a thesis entitled 'Effective Food Supply Chains; generating, modelling and evaluating supply chain scenarios', both at Wageningen University. His research activities focus on Supply Chain Management, Efficient Consumer Response, Logistics Management, Performance Measurement Systems and Tracking and Tracing in food and agribusiness. His research is published in international journals, books and at international conferences. Between 2001-2005 Jack was also active as a senior management consultant for food industries and agribusiness at the consultancy firm Rijnconsult in the Netherlands. He is a fellow of the Mansholt Graduate School. His current research focuses especially on the development of innovative logistics concepts in food supply chain networks and the quantitative modelling and evaluation of such concepts, especially via simulation methodology and applications.

Printed in the United States
by Baker & Taylor Publisher Services